Zdenko Rengel, PhD
Editor

Nutrient Use in Crop Production

Nutrient Use in Crop Production has been co-published simultaneously as *Journal of Crop Production,* Volume 1, Number 2 (#2) 1998.

*Pre-publication
REVIEWS,
COMMENTARIES,
EVALUATIONS . . .*

"**T**his book is an overview on the use of fertilizers in agriculture from various viewpoints: The economy of fertilizer use, their role in global food production, the manufacturer and the farmer, sustainable soil fertility, and environmental impacts. In addition, the unique role of selected nutrients is also reviewed. The broad approach makes the book useful for a wide range of users: Environmentalist, farmers, soil scientists and agronomists, students of life sciences, agronomy, nutrition, ecology and environmental quality, and many other related expertises. The uniqueness of this book is in its broad perspective, which makes it beneficial to a wide range of readers. It is mainly for orientation, rather than a text book. Yet, it treats the basics of fertilizer use in agriculture in its broadest manner."

Moshe Silberbush, PhD
*Head of the Soil & Plant Lab
J. Blaustein Institute for Desert Research
Ben-Gurion University of the Negev
84990 Israel*

"**S**ince world population is increasing, ways must be found to increase crop production without increasing pollution of land, ground water, streams, lakes, and oceans. This will require increased use of added nutrients to increase plant production. It will also be essential to increase the efficient use of added plant nutrients so that environmental pollution can be minimized. This book deals with state of the art efforts to solve these and similar problems. Included are discussions of how to increase the efficient use of plant nutrients and water. Efficient use of water in plant production is becoming more and more important not only because crops require water for plant growth, but also because increasing populations increase non-agricultural demands for water.

Some of the suggested ways of increasing the efficiency of nutrient absorption by plants are coating seeds with fertilizers, managing carefully time and placement of fertilizers, improving water use efficiently, and using plant species and genotypes that are more efficient in absorbing both nutrients and water. This book contains chapters dealing with both micro- and macro-plant nutrients, as well as the effects of below and above

optimum soil pH on the availability of nutrients to plants."

David L. Grunes, PhD in Soil Science
Soil Scientist (Collaborator)
U.S. Plant, Soil, and Nutrition
Laboratory
Tower Road
Ithaca, NY 14853-2091

More pre-publication
REVIEWS, COMMENTARIES, EVALUATIONS . . .

"It is very refreshing to find an agronomy text in which authors do not fight shy of mentioning the policy issues and the agro-politics which so often frustrate their work and their best intentions.

The book begins with an authoritative chapter, on world population, food requirements and fertilizer needs, by B.H. Byrne and B.L. Bumb who coolly sketch the mountain we have to climb. Most of the contributors to the book have been prepared to look out from their laboratory windows and contemplate this awesome prospect. In most . . . chapters *there is a real attempt to suggest how the things we already know could be applied if only the political will and organization could be found.* The time scale is now alarmingly short before greater and sustainable harvests can be achieved, so this is the correct emphasis. . . . In all continents and throughout history, soil fertility has been degraded in farming systems which take more nutrients from the soil than are replaced by natural nutrient cycling or by fertilizers and manures. L.C. Campbell documents these unhappy facts and defines the scale of the problem of food production in terms of the fertilizer requirements. *Both of these opening chapters . . . are extremely readable and get the*

book off to a splendid start. They also provide long and useful lists of references which *provides the reader, unfamiliar with some of the material discussed, with an good jumping-off point for further exploration of the issues. This is a good feature throughout the book, the sources of reference being up-to-date and* (as far as I can judge) *well chosen.*

. . .The subsequent chapters turn to the perennial questions about economical use of fertilizers such as "how much?", "How to deliver?" and "when?".

The last two chapters introduce a genetical perspective and highlight the inexplicable neglect of the development and exploitation of water- and nutrient-efficient genotypes as a way of mitigating problems in the supply of soil resources. . . . the book ends as it began with a reminder that the agronomist and soil scientist cannot be expected to conquer the mountain alone. . . . *this book raises immensely important issues and makes sensible suggestions about where research and agricultural extension work needs to be focused. It is well written and Dr. Rengel has exercised good editorial control."*

David Clarkson, Professor
Department of Agricultural Sciences
AFRC Institute Arable Crops Research
University of Bristol, United Kingdom

Nutrient Use
in Crop Production

Nutrient Use in Crop Production has been co-published simultaneously as *Journal of Crop Production*, Volume 1, Number 2 (#2) 1998.

The *Journal of Crop Production* Monographs/"Separates"

Crop Sciences: Recent Advances, edited by Amarjit S. Basra
Nutrient Use in Crop Production, edited by Zdenko Rengel

These books were published simultaneously as special thematic issues of *Journal of Crop Production* and are available bound separately. Visit Haworth's website at http://www.haworthpressinc.com to search our online catalog for complete tables of contents and ordering information for these and other publications. Or call 1-800-HAWORTH (outside US/Canada: 607-722-5857), Fax: 1-800-895-0582 (outside US/Canada: 607-771-0012), or e-mail getinfo@haworthpressinc.com

Nutrient Use
in Crop Production

Zdenko Rengel, PhD
Editor

Nutrient Use in Crop Production has been co-published simultaneously as *Journal of Crop Production*, Volume 1, Number 2 (#2) 1998.

Food Products Press
An Imprint of
The Haworth Press, Inc.
New York • London

Published by

Food Products Press, 10 Alice Street, Binghamton, NY 13904-1580

Food Products Press is an imprint of The Haworth Press, Inc., 10 Alice Street, Binghamton, NY 13904-1580 USA.

Nutrient Use in Crop Production has been co-published simultaneously as *Journal of Crop Production*, Volume 1, Number 2 (#2) 1998.

Cover design by Thomas J. Mayshock Jr.

Library of Congress Cataloging-in-Publication Data

Nutrient use in crop production/Zdenko Rengel, editor.
    p. cm.
    Co-published simultaneously as Journal of crop production, volume 1, number 2 (#2), 1998.
    Includes bibliographical references and index.
    ISBN 1-56022-061-9 (alk. paper)
    1. Fertilizers. 2. Crops–Nutrition. 3. Soil fertility.
    I. Rengel, Zdenko. II. Journal of crop production.
S633.N88 1998
631.8–dc21
                                       98-39020
                                       CIP

INDEXING & ABSTRACTING

Contributions to this publication are selectively indexed or abstracted in print, electronic, online, or CD-ROM version(s) of the reference tools and information services listed below. This list is current as of the copyright date of this publication. See the end of this section for additional notes.

- *AGRICOLA Database*, National Agricultural Library, 10301 Baltimore Boulevard, Room 002, Beltsville, MD 20705

- *Chemical Abstracts*, 2540 Olentangy River Road, Columbus, OH 43210

- *CNPIEC Reference Guide: Chinese National Directory of Foreign Periodicals*, P.O. Box 88, Beijing, People's Republic of China

- *Crop Physiology Abstracts*, c/o CAB International/CAB ACCESS . . . available in print, diskettes updated weekly, and on INTERNET. Providing full bibliographic listings, author affiliation, augmented keyword searching, CAB International, P.O. Box 100, Wallingford, Oxon OX10 8DE, UK

- *Derwent Crop Production File*, Derwent Information Limited, Derwent House, 14 Great Queen Street, London WC2B 5DF, England

- *Environment Abstracts*, Congressional Information Service, Inc., 4520 East-West Highway, Suite 800, Bethesda, MD 20814-3389

- *Field Crop Abstracts*, c/o CAB International/CAB ACCESS . . . available in print, diskettes updated weekly, and on INTERNET. Providing full bibliographic listings, author affiliation, augmented keyword searching, CAB International, P.O. Box 100, Wallingford, Oxon OX10 8DE, UK

(continued)

- *Foods Adlibra,* Foods Adlibra Publications, 9000 Plymouth Avenue North, Minneapolis, MN 55427

- *Grasslands & Forage Abstracts,* c/o CAB International/CAB ACCESS . . . available in print, diskettes updated weekly, and on INTERNET. Providing full bibliographic listings, author afilliation, augmented keyword searching, CAB International, P.O. Box 100, Wallingford Oxon 0X10 8DE, UK

- *Plant Breeding Abstracts,* c/o CAB International/CAB ACCESS . . . available in print, diskettes updated weekly, and on INTERNET. Providing full bibliographic listings, author afilliation, augmented keyword searching, CAB International, P.O. Box 100, Wallingford Oxon 0X10 8DE, UK

- *Referativnyi Zhurnal (Abstracts Journal of the All-Russian Institute of Scientific and Technical Information),* 20 Usievich Street, Moscow 125219, Russia

- *Seed Abstracts,* c/o CAB International/CAB ACCESS . . . available in print, diskettes updated weekly, and on INTERNET. Providing full bibliographic listings, author afilliation, augmented keyword searching, CAB International, P.O. Box 100, Wallingford Oxon 0X10 8DE, UK

- *Soils & Fertilizers Abstracts,* c/o CAB International/CAB ACCESS . . . available in print, diskettes updated weekly, and on INTERNET. Providing full bibliographic listings, author afilliation, augmented keyword searching, CAB International, P.O. Box 100, Wallingford Oxon 0X10 8DE, UK

- *Weed Abstracts,* c/o CAB International/CAB ACCESS . . . available in print, diskettes updated weekly, and on INTERNET. Providing full bibliographic listings, author afilliation, augmented keyword searching, CAB International, P.O. Box 100, Wallingford Oxon 0X10 8DE, UK

(continued)

SPECIAL BIBLIOGRAPHIC NOTES

related to special journal issues (separates)
and indexing/abstracting

☐ indexing/abstracting services in this list will also cover material in any "separate" that is co-published simultaneously with Haworth's special thematic journal issue or DocuSerial. Indexing/abstracting usually covers material at the article/chapter level.

☐ monographic co-editions are intended for either non-subscribers or libraries which intend to purchase a second copy for their circulating collections.

☐ monographic co-editions are reported to all jobbers/wholesalers/approval plans. The source journal is listed as the "series" to assist the prevention of duplicate purchasing in the same manner utilized for books-in-series.

☐ to facilitate user/access services all indexing/abstracting services are encouraged to utilize the co-indexing entry note indicated at the bottom of the first page of each article/chapter/contribution.

☐ this is intended to assist a library user of any reference tool (whether print, electronic, online, or CD-ROM) to locate the monographic version if the library has purchased this version but not a subscription to the source journal.

☐ individual articles/chapters in any Haworth publication are also available through the Haworth Document Delivery Service (HDDS).

Nutrient Use in Crop Production

CONTENTS

ABOUT THE EDITOR

Zdenko (Zed) Rengel, PhD, is Senior Lecturer at The University of Western Australia. Dr. Rengel has 14 years of teaching and research experience in various aspects of plant physiology, soil fertility, and plant nutrition. He has been awarded four prizes and nine Fellowships from institutions in Austria, France, Germany, the United Kingdom, Japan, the United States, and Australia. He is a consultant to various international and Australian institutions and universities. He has been the invited keynote speaker at eight international conferences and is currently serving on the editorial boards of three international journals. He has published 93 peer-reviewed articles and 10 invited book chapters. His main teaching and research interests are the following: Physiology and genetics of nutrient uptake and transport; soil acidity; aluminium and heavy metal toxicity; land rehabilitation; and modelling of root growth and nutrient uptake.

Preface

Food production by the year 2020 will have to increase about 50% on top of the present levels to satisfy needs of around 8 billion people estimated to be on the Earth by that time. Most of the increase would have to come from intensification of agricultural production. Judicious nutrient management, including fertilization, has been critical in increasing food production to present levels, and will be essential for maintaining soil fertility and food production in the future. Judicious nutrient management balances amounts added to the production system with attempts to maximize nutrient utilization and minimize nutrient losses.

This book summarizes various aspects of optimal nutrient use in modern crop production. The first two chapters deal with issues of production of food, feed and fibre, and their distribution in terms of population demands on one side and soil fertility decline on the other. The third chapter emphasizes the importance of testing soils and plants for the nutrient status to achieve optimal fertilization. Four chapters that follow describe various fertilizer and other nutrient sources, their chemistry and agronomic effectiveness in optimising crop production. Mineral and symbiotic sources of nitrogen are discussed, followed by phosphate and micronutrient fertilizers. Chapter eight evaluates strategies of applying fertilizers through seed soaking and coating, as fertilizer strategies with important niche applications, especially in micronutrient-deficient soils. Chapter nine describes nutrient management strategies, including the role of better adapted cultivars, in optimising water use efficiency, bearing in mind sub-optimal rainfall (amount and/or seasonal distribution) as well as scarcity of quality irrigation water in a major part of arable land in the world. Finally, the last chapter discusses the role of nutrient-efficient genotypes (those that are superior in taking up and utilizing nutrients when grown in nutrient-poor conditions) in modern agricultural production.

I have reviewed all chapters. In addition, I would like to thank Richard Bell (Murdoch University, Perth, Australia), Nick Uren (La

Trobe University, Melbourne, Australia), Richard Richards (CSIRO, Canberra, Australia), Bill Bowden (Agriculture WA, Perth, Australia), and Fusuo Zhang (China Agricultural University, Beijing, P.R. China) for reviewing selected chapters and offering valuable comments and suggestions to the authors.

I would like to express my thanks to the Editor, Amarjit S. Basra, for advice and encouragement, and the staff of the Haworth Press for care and diligence in producing this book.

Zdenko Rengel

Population Growth, Food Production and Nutrient Requirements

Bernard H. Byrnes
Balu L. Bumb

SUMMARY. World population has risen at a rate of 1.9% per year since 1960, but food production has grown at 2.8% per year due to the application of better crop production techniques. Most of the future population growth will occur in developing countries, those with limited ability to feed their growing populations or import food. Fertilizer use to increase production and maintain soil fertility has been essential to increasing food production, and will be essential in the future. World food grain reserves in 1996 were at their lowest levels since the early 1970s, and the rate of increase of food production has slowed. By the year 2020, the population is expected to be 8 billion people. To feed this population, the food grain production will have to increase from the current level of about 2 billion tonnes per year to over 3 billion tonnes. To achieve this level of crop output, intensification of the output on existing land must account for most of the growth, and the amount of fertilizer use will need to increase from 123 million tonnes of nutrients in 1994/95 to over 300 million tonnes in 2020. This requires substantial increase in fertilizer production capacity, which will only occur if relatively stable agricultural markets are established in the countries with expanding populations. The situation in Africa is particularly difficult, with poor input and output markets, declining yield levels due to the

Bernard H. Byrnes, Scientist–Soil Fertility, and Balu L. Bumb, Senior Scientist–Economics, Research and Development Division, International Fertilizer Development Center (IFDC), Muscle Shoals, AL 35662 USA.

Address correspondence to: Bernard H. Byrnes, Research and Development Division, IFDC, P.O. Box 2040, Muscle Shoals, AL 35662 USA (E-mail: bbyrnes@IFDC. org).

[Haworth co-indexing entry note]: "Population Growth, Food Production and Nutrient Requirements." Byrnes, Bernard H., and Balu L. Bumb. Co-published simultaneously in *Journal of Crop Production* (Food Products Press, an imprint of The Haworth Press, Inc.) Vol. 1, No. 2 (#2), 1998, pp. 1-27; and: *Nutrient Use in Crop Production* (ed: Zdenko Rengel) Food Products Press, an imprint of The Haworth Press, Inc., 1998, pp. 1-27. Single or multiple copies of this article are available for a fee from The Haworth Document Delivery Service [1-800-342-9678, 9:00 a.m. - 5:00 p.m. (EST). E-mail address: getinfo@haworthpressinc. com].

lack of nutrients, and continued population growth; there are few indications that fertilizer use will soon increase to reduce the rate of soil degradation and to produce the needed food. *[Article copies available for a fee from The Haworth Document Delivery Service: 1-800-342-9678. E-mail address: getinfo@haworthpressinc.com]*

KEYWORDS. Fertilizer needs, food production, nutrients, world population

INTRODUCTION

Humanity's future is at a critical juncture as the population continues to grow at a high rate in the most food-deficit regions of the world, while world grain reserves are at their lowest levels in 20 years. There is little land area that can be readily brought under economic agricultural production, additional land suitable for irrigation is limited and water is also limited. With excess food production in developed countries, and the success of increasing grain production in many developing countries, especially India and China, there has been a reduced interest in increasing the food production capabilities of the developing countries. In addition, fertilizers are scrutinized by some as sources of pollution rather than a means to produce more food and provide food security for the developing countries.

The movement of nutrients through ecosystems is a natural process, but the rates of movement are greatly increased when land is used for agricultural purposes, as crop produce moves from fields to places of consumption, in recent times the rapidly growing urban areas. In addition, soil tillage causes more rapid degradation of organic material, loss of nutrients, and increased soil erosion. The most economical way to replenish nutrients is normally with mineral fertilizers, but the utilization of locally available organic materials and recycling of nutrients in crop residues are very important to maintaining long-term fertility and productivity of soils, particularly those of the tropics, where food production must increase in the future. Soil productivity includes all factors which affect plant growth, including soil fertility, water availability, climate, etc.

This chapter analyzes the trends in population growth and food requirements, and the role that fertilizers have played and will continue to play in meeting food and fiber requirements for the world's growing population.

POPULATION GROWTH, 1950-2025

At the turn of the century, the world population was estimated to be less than 2 billion (Figure 1). The population reached 2.5 billion in 1950, and 3 billion in 1960 (Bongaarts, 1994). The world population began to increase even more rapidly after 1960, due to continued high birth rates with decreasing death rates brought about by improved medical care and disease control. Many developing countries had 2-3% annual growth rates of their populations. Consequently, the world's population more than doubled between 1950 and 1995. The world population in 1995 was estimated to be 5.7 billion people and is expected to reach 6.3 billion by the end of this millennium. From 1995 to 2025, the world population was expected to increase by 1.4% per year, to 8.5 billion people. Most analysts believe that after 2025 the population growth will slow and the population will stabilize at 11-12 billion in 2150 (UN, 1992). The most recent projections of the UN Population Division are that the population will only reach 8 billion in 2025.

More than 95% of the anticipated population growth will occur in the developing regions of Asia, Africa, and Latin America (UN, 1992). The combined population of these regions (identified as Group II in Table 1) will increase from 4.6 billion people in 1995 to 7.3 billion people in 2025. During the next 30 years, the population is projected to more than double in Africa, and to increase by 45% in

FIGURE 1. World population over time. Modified from Bongaarts (1994); data from the Population Reference Bureau (1995), and UN Population Division.

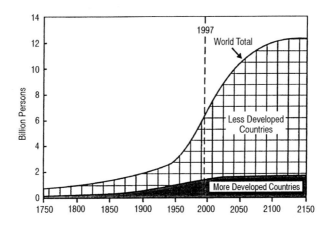

TABLE 1. World population in various regions

Region	1950		1995		2000		2025	
	(million)	(%)	(million)	(%)	(million)	(%)	(million)	(%)
Group I	752	29.9	1,121	19.7	1,143	18.3	1,237	14.5
North America	166	6.6	293	5.1	295	4.7	332	3.9
Europe	393	15.6	507	8.9	510	8.1	515	6.1
Eurasia	180	7.1	293	5.1	308	4.9	352	4.1
Oceania	13	0.5	28	0.5	30	0.5	38	0.4
Group II	1,766	70.1	4,581	80.3	5,118	81.7	7,267	85.5
Africa	222	8.8	720	12.6	867	13.8	1,597	18.8
Latin America	166	6.6	481	8.4	538	8.6	757	8.9
Asia	1,378	54.7	3,380	59.3	3,713	59.3	4,912	57.8
World	2,518	100.0	5,702	100.0	6,261	100.0	8,504	100.0

Note: Due to rounding, totals may not add up.
Sources: United Nations (1992), except the 1995 data that are from Population Reference Bureau (1995).

Asia. In absolute numbers, Asia will add more people to its existing population than Africa because Asia's population is already over 3 billion, compared to 720 million in Africa. In Latin America, the population will increase by 57%. Overall, the developing regions will account for almost 86% of the global population in 2025. Obviously these regions, which already have food deficits, face the greatest challenges in feeding an additional 2.7 billion people–over 10 times the current population of the U.S.A. In contrast to Group II, the countries of Group I are expected to have only a modest increase in population, 116 million persons, or less than one-twentieth of the projected increase in Group II.

Although estimates vary, depending on the consumption level used, a reasonable estimate is that food production will have to increase by over 50% between 1995 and 2020 to feed the expected population. The developed countries will have little problem in producing enough food and fiber for their populations, but the large increases in the populations of Africa, Asia, and Latin America will place tremendous stress not only on the production capabilities of those regions, but also of the world.

It is clear that the problem is not simply a production capacity problem, but a food availability and distribution problem. One estimate concludes that even with a low level of technology, total food production by the 117 developing countries would be enough to feed up to 50% higher population than they are expected to have in the year 2000, but the other 64 countries will not be able to feed their populations of 1.1 billion (WCED, 1987). Since many of the developing regions will not be self-sufficient, they will have to depend on imports to meet their food requirements, some of it will likely be humanitarian aid. Rosegrant, Agcaoili-Sombilla, and Perez (1995) estimate that the developing regions will need to import about 188 million tonnes of cereals and 15 million tonnes of soybeans by 2020.

The reforming regions (Eastern Europe and Eurasia) have had significant decreases in their food production, but as they recover through political and economic reforms, they may become important exporters of grain.

Food production requires fertilizers, either organic or manufactured, to replace nutrients removed in harvested crops and to increase the yield potential. Continued fertilizer use will be required to sustain food production in the developed countries, while increased use will be required in many of the developing countries to increase food production.

HISTORIC TRENDS IN FOOD
AND FIBER REQUIREMENTS

Increased population not only increases the demand for food and fiber, but also reduces per capita area of land available for food and fiber production. In the late 18th century, during an era of slow scientific and technological advancement, Malthus (1798) espoused that the balance of stagnant food supply and rapidly increasing food demands could only be restored by reducing the population through famine, disease, and war. More modern analysts have also predicted mass starvation and civil unrest caused by population growth and limited food supplies. Paddock and Paddock (1967) and others predicted mass starvation in Asia and Africa. However, with only limited application of science-based technologies to agriculture, food production increased at a higher rate than the population from 1961-90. In the mid-1990s, analysts again assert that the "carrying capacity" of the Earth will soon be exceeded, and that feeding an additional 2.7 billion in the next 30 years is impossible (Brown and Kane, 1994). Some of the agencies which are credited with the development of technologies which have enabled the developing countries to meet their food needs now question whether wider adoption of these technologies is desirable, and they have shifted away from efforts to increase food production, to controlling population and addressing environmental/ecological problems (Easterbrook, 1997). Others believe that with the application of known technologies, and most of all, political will, the world will be able to feed about 8 billion people in 2020 (Seckler, 1994; IFPRI, 1995; Harris, 1996), and that population growth will slow only when poverty has been reduced.

The increased food and fiber production in the developing world, particularly Asia (the "Green Revolution"), was brought about by the development and adoption of improved crop varieties, irrigation, fertilizers, and crop protection chemicals. From 1961 through 1990, world food production increased at an annual rate of 2.8%, while the population increased at 1.9%. This increase in production induced a long-term decline in crop prices, which helped poor people to buy food at lower costs. However, in some countries, food prices are kept low by taxing agriculture to placate urban consumers (Zaman, 1991). Such a policy is counterproductive in the long term.

In absolute numbers, world cereal production increased from 876 million tonnes in 1961 to 1,950 million tonnes in 1990 (Table 2; see also FAO, 1996b). Because cereals provide 60-70% of the average per

TABLE 2. Projected cereal demand and yield needed to achieve it

Crop		Current Production	Projected Demand		Yield Required		
		(million tonnes)			(tonne/ha)		
	Year	1990	2000	2025	1990	2000	2025
Wheat		600	740	1,200	2.4	2.8	4.4
Rice		520	640	1,030	2.4	3.1	5.3
Maize		480	620	1,070	3.7	4.1	5.8
Barley		180	220	350	2.3	2.7	4.1
Sorghum/Millet		85	110	180	1.5	1.8	2.6
Total Production[a]		1,950[b]	2,450	3,970			

Source: Borlaug and Dowswell (1993).
[a]Includes other cereals not listed (oats, buckwheat, rye, etc.).
[b]The latest FAO estimate for 1990 was reduced from earlier estimates of 1,970 million tonnes.

capita caloric intake, they are treated as an indicator of food production. In addition, production of tubers, fruits, vegetables, oil seeds, and fibers also increased significantly and will need to increase in the future. Despite the unprecedented growth in food production over the last 25 years, about 786 million people in the developing countries currently suffer from hunger and malnutrition, and 40,000 people die daily as a result of poor nutrition (Speth, 1993). The number of food-deficit countries in the various regions of the world has increased only slightly since the mid-1960s, except for Africa. The number of food-deficit countries has grown from 28 to 41 in sub-Saharan Africa, where one out of every four persons is thought to be food insecure (IFAD, 1993).

The International Food Policy Research Institute (IFPRI) has estimated that to meet the food requirements of the projected population in 2020, annual cereal production will need to increase by 57%, from the current 1,950 million tonnes to an estimated 3,066 million tonnes (Rosegrant, Agcaoili-Sombilla, and Perez, 1995). Harris (1996) projected a need for 3,143 million tonnes in 2025. In addition, production of other food items would need to increase similarly.

IFPRI bases its food requirement projections on the incomes of the national populations. By using average incomes, it likely underestimates the food requirements of the people living in poverty, especially those earning less than one US dollar/day (Pinstrup-Andersen and Pandya-Lorch, 1994). According to World Bank (1992) projections, about 50% of sub-Saharan Africa's population and 37% of South Asia's population will be living in poverty in the year 2000. If adequate nutrition is to be provided to the people living in poverty, an additional 400 million tonnes of cereals will be needed (Hazell, 1994). Borlaug and Dowswell (1993) estimated that world cereal production should increase from 1,950 million tonnes in 1990 to 3,970 million tonnes in 2025 (Table 2). Based on these estimates, world production of cereals will need to increase by 57-103% to meet the food requirements of 8.0-8.5 billion people in 2020-2025. Similarly, production of non-cereal food and fiber crops should increase by 73-82% (Table 3).

THE ROLE OF FERTILIZERS
IN MEETING FOOD REQUIREMENTS

The rates of nutrient movement are greatly increased in any type of commercial agriculture. While subsistence agriculture may entail small

TABLE 3. World production of non-cereal crops (million metric tonnes)

Crop	1969-71 Average	1979-81 Average	1989-91 Average	% Annual Growth[a]	2020 Projection	% Annual Growth (1990-2020)
Total non-cereal crops	2,030	2,498	3,065	2.0	5,296	1.8
Roots and tubers	539	548	575	0.3	798	1.1
Pulses	42	41	56	1.4	96	1.8
Oilseeds	97	150	199	3.6	360	2.0
Sugar (cane and beet)	799	1,042	1,354	2.6	2,453	2.0
Vegetables	270	363	457	2.6	828	2.0
Fruits	237	299	355	2.0	643	2.0
Tree crops	33	39	48	1.9	80	1.7
Beverages and nuts	13	16	21	2.4	38	2.0
Fibers	41	48	60	1.9	109	2.0

Source: FAO (1996a) for actual data supplemented with authors' projections.
[a]1969-71 to 1989-91.

amounts of movement, commercial agriculture entails much more movement, as products are produced in one location, harvested, and then consumed in another location. In addition, nutrient cycles are dramatically changed by clearing the land and cultivating the soil. Organic matter turnover and soil erosion are greatly increased. Since the cultivation of crops began, farmers have strived to replenish nutrients removed, both in harvested crops and soil erosion, to increase, or at least maintain, crop production. Early farmers could only relocate organically-derived nutrients from animal or plant manure and utilize slash-and-burn practices to replenish nutrients to the surface soil. The failure to maintain soil fertility under increasing population pressures has been the demise of many flourishing societies (Ponting, 1990). Mineral fertilizers have greatly reduced the transportation and labor costs required to replenish and increase nutrient supplies, and have greatly contributed to the increased world food production (Baanante, Bumb, and Thompson, 1989; Pinstrup-Andersen, 1993). While these issues are evident to farmers and agricultural scientists, the policy-makers and the public may not understand that fertilizers are needed to replenish the nutrients removed in harvested crops as well as by natural processes.

The increased cereal production in the past 25 years was brought about by both an increase in the area cultivated and an increase in crop yield, but the increase in yield contributed relatively more than the increase in cropping area. Yields of major cereals–wheat, rice, and maize–increased by 50-70%. The higher crop yields played a major role in many regions, including Asia and North America, where the cultivated area decreased, and yet cereal production greatly increased. While improved crop varieties increased the yield potential, the use of fertilizers, along with irrigation and crop protection chemicals, enabled the realization of that potential. Without fertilizers, the high-yielding varieties (HYVs) do not produce higher yields than traditional varieties. Cereal production and fertilizer use are closely related (Figure 2).

Because of complementarity with other inputs and crop production technologies, it is difficult to separate the effect of fertilizers on crop yields from effects of the other inputs necessary for increasing production. Based on the information available from FAO and other sources,

FIGURE 2. Growth in fertilizer use and cereal production in developing countries, 1961-95. From FAO (1996a)

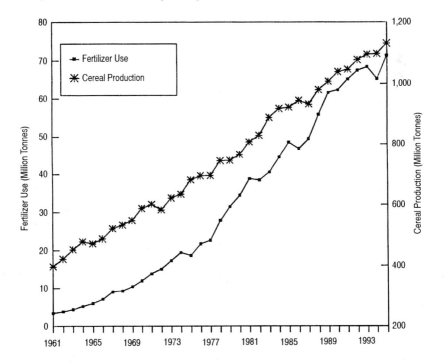

the following assumptions are made to estimate fertilizers' contribution to grain production: (1) 10 tonnes of cereal grain is produced for every ton of N use in developing countries, and 15 tonnes of cereal/ton N use in the developed countries, and (2) 65% in developing countries and 50% of additional N use in developed countries is devoted to grain production. Approximately 476 million tonnes, or 44% of the increase in cereal production during the 1961-90 period can be attributed to increased N use at a global level. In individual countries, especially in Asia, this percentage is likely to be about 60% (Vaidyanathan, 1993). Such estimates, of course, are only suggestive, since yield is not improved with fertilizers without appropriate varieties, weed control, and other production inputs, while without fertilizers, yields decline to very low levels, even with the other inputs.

Annual global fertilizer use increased from 27 million nutrient tonnes in 1959/60 to 143 million nutrient tonnes in 1989/90 (Table 4). A year's consumption is expressed as two calendar years because fertilizer

TABLE 4. Fertilizer use by nutrients and regions, 1959/60-1994/95

Year/Nutrient	North America	Western Europe	Eastern Europe	Eurasia	Oceania	Africa	Latin America	Asia	World
				(million nutrient metric tonnes)					
1959/60									
N	2.55	3.26	0.60	0.71	0.03	0.22	0.29	1.87	9.54
P_2O_5	2.56	3.72	0.55	0.84	0.73	0.23	0.30	0.82	9.75
K_2O	2.04	3.76	0.61	0.79	0.07	0.08	0.14	0.64	8.13
Total	7.14	10.74	1.76	2.34	0.84	0.53	0.73	3.34	27.42
1989/90									
N	11.25	11.17	4.56	9.92	0.50	2.04	3.80	35.91	79.14
P_2O_5	4.55	5.08	2.29	8.17	1.05	1.08	2.40	12.76	37.39
K_2O	5.08	5.76	2.39	6.38	0.26	0.49	2.05	4.47	26.89
Total	20.88	22.00	9.25[a]	24.47[a]	1.81	3.61	8.26	53.14	143.42
1994/95									
N	11.94	93.78	2.03	2.80	0.72	2.02	3.96	40.35	73.60
P_2O_5	4.62	3.79	0.59	0.71	1.25	1.03	2.88	14.79	29.66
K_2O	5.00	4.26	0.50	1.05	0.34	0.48	2.45	5.96	19.99
Total	21.55	17.83	3.10	4.56	2.30	3.53	9.29	61.09	123.25

[a]Before the political and economic reforms, fertilizer use reached 10.11 and 27.19 million nutrient metric tonnes in 1988/89 in Eastern Europe and Eurasia, respectively.
Note: Due to rounding, totals may not add up. Source: FAO (1996b).

statistics are reported on a July through the following June basis. Of the three fertilizer nutrients, N has shown the largest absolute and relative growth, increasing from 10 million nutrient tonnes in 1959/60 to 79 million nutrient tonnes in 1989/90. Economic and political reforms in the former Communist countries caused fertilizer use to decrease drastically in Eastern Europe and Eurasia since 1988/89. Consequently, global N use decreased from 79 million nutrient tonnes in 1988/89 to 74 million nutrient tonnes in 1994/95. Grain surpluses caused low crop prices and reduced N use in North America and other developed regions in the 1980s. However, in the 1990's, as grain supplies decreased, increases in grain prices brought about increased N use and food grain production.

The rate of N use in Asia increased in the 1970s and the 1980s. With responsive HYVs of rice and wheat, N use rose from 6.5 million nutrient tonnes in 1970 to over 40 million tonnes in 1994/95. The higher use of N caused higher yields and more P and K offtake and use (Table 4). In Eastern Europe and Eurasia, P_2O_5 and K_2O use decreased after 1988/89. The reduced consumption in these regions caused a decrease in P_2O_5 and K_2O use at the global level between 1988/1989 and 1993/1994 (Bumb and Baanante, 1996a).

The Case of Africa

Increasing fertilizer use and resulting increased food production are occurring in most of Latin America and Asia. In addition, the rate of population growth is actually less than the demographers had earlier projected. The situation in Africa, however, presents more intransigent problems in all three of these concerns. The population of sub-Saharan Africa is projected to double in the next 25 years, and per capita food production has been declining. Current yields of 1 ton/ha will need to triple by 2025 to feed the projected population (Harris, 1996).

The low levels of fertilizer use in much of Africa is a major concern because the fertility of the very nutrient-deficient soils continues to decline. While there are some productive soils in the central part of the continent, as well as other places, the soils of much of the rest of the continent, particularly West Africa, are highly weathered and thus very infertile. While the soils can become productive in all but the most arid regions by increasing the soil fertility in respect to almost all nutrients and by increasing the soil pH (Blackie, 1995), it is difficult to manage nutrients on these soils because they have low cation exchange capacities, low organic matter contents, and many are highly leached, which makes retention of nutrients difficult. Nutrient levels cannot be raised to the levels of many temperate soils without high losses from the systems.

Fertilizer use averages only about 9 kg per ha on the arable land in sub-Saharan Africa (Table 5), and most of the use is on export crops rather than food grain crops. An exception is Nigeria, which uses more fertilizer, especially more on grain crops, than the other countries because of fertilizer subsidies (IFDC, 1996).

Farmers in many areas simply lack the economic incentives and the opportunities to use fertilizers profitably; the value:cost ratio of additional grain production to fertilizers is not favorable. Fertilizer prices at local levels (Table 6) are often much higher than the "world market price" due to currency devaluations, lack of fertilizer storage, transportation, and marketing capabilities. At the farm level, high risk due to weather, poor access to crop output markets and credit, as well as unfavorable prices at harvest, greatly constrain fertilizer use on food grains in most of West and Central Africa (Gerner and Harris, 1993). Local grain prices often fluctuate by a factor of 10 during the year.

TABLE 5. Regional fertilizer use per hectare in 1994/1995

Region	kg $N-P_2O_5-K_2O$/ha
Africa	19
Sub-Saharan Countries	9
Republic of South Africa	63
North African Countries	51
Asia	129
West Asia	60
South Asia	77
East Asia	216
South America	66
Central America	62
North America	92
Eurasia (Former Soviet Countries)	20
Oceania	44
Western Europe	201
Eastern Europe	67
World	85

Source: FAO (1996a, b).

TABLE 6. Comparisons of nitrogen fertilizer prices in various countries between 1985 and 1990

| Country | US $ Per Tonne N | | % Change | |
	1985	1990	In US $	In Local Currency
Bangladesh	386	335	− 15	+ 7
Turkey	216	260	+ 17	+ 389
Ghana	505	895	+ 44	+ 697
Zambia	1250	1294	+ 3	+ 1,384
Mexico	164	177	+ 7	+ 975
Venezuela	188	96	− 95	+ 819
Poland	179	685	+ 74	+ 3,648
United States	275	284	+ 3	+ 3

Source: Bumb and Baanante (1996a).

Even before the tremendous increase in fertilizer prices at a local level, fertilizer use on maize was only profitable in Ghana, for instance, if the farmer stored the grain for 7.5 months (Pouzet, 1991).

In addition to the economic constraints and risks that inhibit fertilizer adoption, land tenure, and poor government policies also constrain use (Breman, 1995). Farmers lack incentives to make long-term improvements, not only related to soil fertility or liming materials, but also for the adoption of appropriate crop rotations. Without land tenure, they lack incentives to plant long-duration crops such as trees, improve the soil pH and organic matter contents, and control erosion. Lack of land title also prevents farmers from borrowing money for fertilizers and other inputs and from investing in land improvements to conserve water and reduce erosion.

The low fertilizer use in sub-Saharan Africa is causing extreme nutrient depletion of the very infertile soils (Smaling, 1993; 1995; Vlek, 1993; Van Reuler, 1996). Estimates of nutrient balances on a micro scale (Table 7) translate to a rate of nutrient depletion in sub-Saharan Africa of approximately 8 million tonnes per year in 1983, which will increase to 13 million tonnes per year in 2000 (Stoorvogel and Smaling, 1990). Under the circumstances just described, the increased and well-managed use of fertilizers has little opportunity to replace the low external input systems which are rapidly degrading the productivity of the land and will have a huge cost to the populations

TABLE 7. Estimated annual negative nutrient balances of sub-Saharan African countries

Country	Net Nutrient Loss		
	N	P	K
	(kg/ha/year)		
Benin	14	1	9
Botswana	0	(+1)	0
Cameroon	20	2	12
Ethiopia	41	6	26
Ghana	30	3	17
Kenya	42	3	29
Malawi	68	10	44
Mali	8	1	7
Nigeria	34	4	24
Rwanda	54	9	47
Senegal	12	2	10
Tanzania	27	4	18
Zimbabwe	31	2	22

Source: Smaling (1993).

and the environment. There has been a large amount of effort to promote fertilizer use, but the root cause for the lack of adoption is not addressed; farmers lack economic incentives to use fertilizers, due to poor input and output marketing systems, poor infrastructure for transportation and storage of produce, output taxation, and lack of land tenure.

Dr. Norman Borlaug, who received a Nobel Prize for his work on developing HYVs of semi-dwarf wheat, has recently been involved in efforts to improve crop production in sub-Saharan Africa, mainly by addressing the severe nutrient needs (Borlaug and Dowswell, 1993; Easterbrook, 1997). With the Sasakawa Global 2000 Project, he helped demonstrate that grain production on individual farmer's fields in Ghana could be increased three- to four-fold by providing improved seeds, fertilizers, and education on improved agronomic practices. However, without infrastructure to transport, store, and market the additional production, the local price of the harvested grain dropped to almost nothing and the farmers were not able to profit from the additional yields.

Policy initiatives by both the developing and developed countries will be required to reverse the steady decline of the natural resource base of Africa and to help the countries meet their food needs. The development of fertilizer marketing systems has also been severely inhibited by the vagaries of international aid given in the form of fertilizers. In many years, half of the countries of sub-Saharan Africa used no fertilizers except for what was given to them (Gerner and Harris, 1993). This is not conducive to developing the needed fertilizer marketing. While efforts to privatize the fertilizer sectors are currently under way in many developing countries, the industry in Africa will need considerable support, protection from competition with subsidized donated fertilizers, and freedom from government disincentives to develop viable fertilizer distribution systems. Infrastructure improvement and education to foster improved land management and market access are needed. Fertilizer aid needs to be given through government policies that foster development of sound fertilizer use and commodity marketing systems.

Clearly, the extensive, low-input systems of these countries are not sustainable (Larson, 1993; Van der Pol, 1993). The lack of fertilizer use will not change until the economic constraints to use fertilizers are eliminated: (1) security of tenure and cultivation rights must be guaranteed, (2) the farmers develop appropriate skills and have access to services needed to properly utilize fertilizers, and (3) economic benefits through access to input and output markets are available to farmers. Carefully formulated and implemented government policies are needed to bring about adoption of technologies to reverse the environmental and natural resource degradation in Africa and to feed future populations (Hitzhusen, 1993). The decline of the resource base and loss of food self-sufficiency will continue unless both the developed and the developing country governments adopt better policies and programs to address the African problems.

TRENDS IN FERTILIZER PRODUCTS AND USES

The fertilizer industry is becoming a global industry. Less efficient production sites are being largely abandoned, and government-owned and operated plants are being privatized in many cases. There is highly competitive pricing, and effective distribution in areas experiencing growth in use. The fertilizer industry continues to move toward production of higher analysis materials, which is of some benefit for developing countries with poor transportation infrastructure. Unfortunately, the governments of these same countries often lack the ability to provide reliable advice on using fertilizers nor do they effectively monitor fertilizer quality and marketing to ensure fair treatment of farmers.

Nitrogen Products

The production capacity for urea greatly increased since 1970, which coincided with the development and adoption of HYVs of rice that could utilize the higher N availability to produce higher yields under flooded conditions. In 1993/94, urea accounted for over 40% of the global N use (Table 8). Ammonium nitrate and calcium ammonium nitrate accounted for 9% and 5%, respectively, of the global N use. In Eurasia, Western Europe, and Eastern Europe, these products account for 45-58% of N fertilizers.

In North America, large scale and highly mechanized farming has promoted the use of bulk blended and liquid fertilizers, which can be applied accurately to meet the nutrient needs on specific fields. There will be increased bulk blending in many of the developing countries, as they develop the capability to more accurately assess the nutrient needs of individual fields through soil testing. The higher efficiency of

TABLE 8. Market share of N products in the regions of the world, 1993/94

Product	World	North America	Western Europe	Eastern Europe	Eurasia	Oceania	Africa	Latin America	Asia
				(% shares in total)					
Ammonium sulfate	3.3	1.8	2.6	3.7	4.5	1.7	3.8	14.3	2.7
Urea	41.2	17.2	13.1	21.5	18.8	43.2	32.7	50.6	60.0
Ammonia	6.9	35.4	0.4	0	4.7	7.9	0.3	6.8	0.1
Ammonium nitrate	9.1	5.4	17.5	43.5	43.1	2.4	27.4	7.1	2.0
Calcium ammonium nitrate	5.2	0.1	28.2	14.5	2.4	0.6	7.7	1.5	1.7
N solutions	5.2	20.4	9.3	3.2	4.7	0.2	0	0	0.1
Ammonium phosphates	4.2	5.9	0	0.5	4.0	19.2	3.4	6.9	4.5
NP/NK/NPK	11.5	12.4	26.6	13.0	15.4	24.9	21.3	8.0	6.7
Other N products	13.3	1.4	2.4	0	2.4	0	3.4	4.9	22.2

Note: Due to rounding, totals may not add up.
Source: Derived from IFA data (IFA, 1995).

side-dressed ammonium nitrate fertilizers relative to urea will help in the adoption of nitrate fertilizers, particularly on high value crops like vegetables. In Asia, about 40% of N is used on flooded rice, so urea is the dominant fertilizer. In Asia and Latin America, the N fertilizer industry has developed in the 1970s and the 1980s, when the technology and investment costs favored the construction of large ammonia-urea complexes. Outside of North Africa and the Republic of South Africa, fertilizers are mainly imported ammonium sulfate and NPKs. Use of ammonium nitrate and calcium ammonium nitrate (CAN) due to agronomic and soil considerations in Egypt has contributed to a relatively larger share of nitrate-based fertilizers in Africa. These regional use patterns are expected to continue.

Phosphorus Products

Due to previous fertilization practices, many of the soils of North America and Western Europe have high P fertility levels so they will require, principally, maintenance doses of P to replace the P removed in the harvested crop. For this reason, little increase in P fertilizer consumption in the developed countries is expected. However, consumption in the developing countries will have to increase rapidly to meet projected food needs (Engelstad and Hellums, 1993). In the last decade, there have been significant increases in the use of P fertilizers in Asia, Latin America, and North Africa (Schultz, Gregory, and Engelstad, 1993; Bumb and Baanante, 1996b). The prospects for increasing P use in much of sub-Saharan Africa (excluding the Republic of South Africa) remain bleak.

Diammonium phosphate (DAP) (18-46-0) and monoammonium phosphate (MAP) (11-52-0) now accounts for about 30% of the global phosphate consumption. In 1993/94 single superphosphate (SSP) accounted for about 19% and triple superphosphate (TSP) only about 9% of total P use on a nutrient basis (Figure 3). The increase in market share by the ammonium phosphates has been due to shifting production away from triple superphosphate. Single superphosphate (0-20-0-6S) continues to be an important P source in countries with SSP production capacity and in areas with known S deficiencies. While the ammoniated phosphates represent an economic means of addressing both P and N requirements, they do not contain Ca and S found in single superphosphate or many of the NPKs. As more ammoniated phosphates are used, the needs for Ca and S will have to be met by other

FIGURE 3. Total use of various P products on a nutrient basis, 1993/94. International Fertilizer Association (IFA) World Fertilizer Consumption Statistics No. 27, Paris, December 1995.

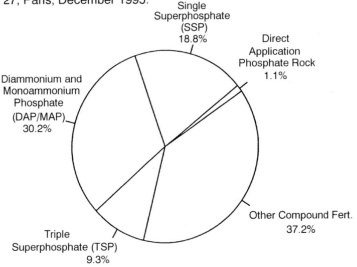

products. On the other hand, they present opportunities for bulk blending with urea and reduced transportation costs on a per nutrient basis that SSP and TSP do not have.

In the past 25 years, there has been considerable effort to determine the economic feasibility of developing phosphate rock (PR) deposits in Africa for direct application. There has also been work to identify cheaper methods of benefication to produce more water-soluble materials. While many tropical soils have characteristics that favor the direct application of PRs (i.e., very low pH and low levels of P and Ca), most of the PRs in Africa have low reactivity and are not effective sources of P for annual crops. For the few PRs which have medium to high reactivity, numerous studies have shown that they can be appropriate alternatives to high analysis P fertilizers under reasonably humid climatic conditions only under very limited conditions (Hammond, Chien, and Mokwunye, 1986). Unfortunately, the costs of mining, processing, and transporting the PRs result in unfavorable economic response for crops compared to responses from imported water-soluble products.

A large increase in the world market price for finished soluble P fertilizers is not expected in the near future. Surplus production capacity in the past resulted in low prices, but now demand has caught up with capacity. A modest increase in prices is necessary to promote further expansion of capacity. Increased production is expected from Morocco, Jordan, Iraq, and South Africa. Presently, 75% of PR production is concentrated in the United States, the Former Soviet Union, Morocco, and China (Van Kauwenbergh, 1995). While PR is the non-renewable resource used to produce all P fertilizers, there are ample supplies of quality PR, enough for hundreds of years at expected rates of consumption.

Research and extension efforts are needed in the developing countries to increase the efficiency of soluble P fertilizer use on upland crops by promoting band application of P, placement with other nutrients, and more use of liming materials. These approaches provide P more efficiently in concentrated zones and an economic means to build P fertility. To foster adoption of these techniques, simple application equipment is needed in the developing countries.

Potassium Products

The need for potassium fertilizers is increasing throughout the world as use of the other nutrients increases the removal of K, and native levels are being more rapidly depleted. For instance, the combined rice and wheat production of South Asia has increased from about 57 million tonnes in 1950 to about 128 million tonnes in 1990. If 30 kg K_2O per tonne of grain is removed in a harvested crop, an additional 2.1 million tonnes of K_2O is removed each year. This is twice the annual current use of K_2O for the region (Bumb and Baanante, 1996a). China is importing large amounts of K, but many of the developing countries have not increased their importation of the nutrient to maintain or, if needed, increase the K status of their soils.

Demonstration and extension work will be needed to help farmers determine if they need more K, particularly under intensive cropping. In the poorest regions, little of the K-containing crop residue remains in the field, greatly increasing the removal of K. Soil testing is needed in many of the developing countries to identify the K deficiencies which often develop erratically in regions since removal varies greatly from field to field. Soil testing and analysis of archival soil test information would help to document these changes and suggest research, extension, and policy initiatives that could help ameliorate the decline in K status of soils in the developing countries.

ROLE OF FERTILIZERS
IN PREVENTING SOIL DEGRADATION

Fertilizers not only increase food production to feed people but also contribute to the preservation of the natural resource base and biodiversity. First, if the nutrients removed in harvested crops and erosion are not replaced, soil fertility declines and soils are degraded, as is happening in much of Africa. Second, application of fertilizers results in more biomass production, which, if left in the field, will maintain the organic matter and soil cover. Third, fertilizers facilitate the adoption of HYVs capable of raising crop yields. By producing more grain output from the same area of cultivated land, pressures to cultivate additional marginal lands and clear forests are reduced (Mellor, 1988; Hitzhusen, 1993; Bumb and Baanante, 1996a). For example, in 1940, the farmers in the US harvested 252 million tonnes of food, feed, and fiber crops by cultivating 129 million ha. In 1990, they produced 600 million tonnes of crops from 119 million ha of land by using 21 million tonnes of fertilizers (as well as other technologies). With the technologies of the 1940s, the United States would have needed an additional 188 million ha (160% more) of forest and grassland to produce the same amount of crop output (Quinones, Borlaug, and Dowswell, 1996). Much of the additional land brought under cultivation would be marginal, which is more risk-prone, and more susceptible to degradation. Thus, increased grain production through the adoption of modern technologies including fertilizers and better management, has saved soils, forests, recreational areas, and biodiversity. Fourth, lime and fertilizers ameliorate very unproductive soils so they can become productive. Addition of phosphate fertilizers and lime has transformed the Cerrados of Brazil into a highly productive area, important to the national economy (Borlaug and Dowswell, 1993). Thus, by preserving and sustaining the natural resource capital of the land and forests, fertilizers and other improved crop production practices help preserve productivity resource and maintain a greater degree of inter-generational equity.

Fertilizer Demand and Requirements in 2020

Fertilizer demand is projected by taking into account current trends and likely changes in economic, technical, and policy variables, whereas fertilizer requirements are estimated by taking into account food production requirements, nutrient use efficiency and losses, and supply of nutrients from non-fertilizer sources (Bumb and Baanante, 1996b). Fertilizer demand is projected to reach 208 million nutrient tonnes by 2020, 86 million nutrient tonnes in developed countries and 122 million nutrient tonnes in developing countries. The demand by nutrients and regions has been estimated (Bumb and Baanante, 1996b) for the year 2020 (Table 9). However, the projected demand will fall short of the nutrient requirements for food production and nutrient replenishment (Table 10). This shortfall will be more severe in the developing countries, jeopardizing the goals of food security and sustainable agriculture.

TABLE 9. Projected fertilizer demand by region and nutrients in 2020

Region	Nitrogen	Phosphate	Potash
	(million nutrient tonnes)		
North America	14.0	6.0	6.0
Western Europe	10.0	4.0	4.5
Eastern Europe	5.0	2.5	2.6
Eurasia	10.5	7.0	6.5
Oceania	1.2	1.5	0.5
Africa	5.4	2.6	1.3
Latin America	7.0	5.0	4.2
Asia	62.2	27.4	11.1
TOTAL	115.3	56.0	36.7

Source: Bumb and Baanante (1996b).

TABLE 10. Projected global fertilizer demand and requirements in 2020

Region	Projected Demand	Estimated Requirements	
		Nutrient Removal Approach	Cereal Production Approach
	(million nutrient tonnes)		
Developed countries	86	115	78
Developing countries	122	251	185
World	208	366	263

Source: Bumb and Baanante (1996b).

OTHER DEVELOPMENTS

In the US, "precision agriculture," or site-specific management, is an emerging crop management technology which some believe will be as revolutionary to crop production as mechanization or HYVs. It involves intensive mapping of soil characteristics and yield monitoring with computer data management and global positioning systems (GPS) to vary fertilizer application rates, pesticide application rates, and seeding densities across fields in response to soil characteristics of small areas of the field. These areas are 0.2-0.5 hectare which are sampled and the fertility status in respect to each nutrient, organic matter, and other parameters are determined, and the data is input into a computer-based geographical information system (GIS). By using computer-controlled systems on the field equipment, variable rates of fertilizer and pesticide application, planting densities, yield monitoring, and other things can be controlled or recorded. In addition to site-specific management, which produces higher yields with minimized inputs, the system provides accurate and detailed record keeping, allows farmers to visualize how yield and fertility vary across fields, and suggests possible reasons for yield variations.

While new technologies to better manage fertilizers emerge periodically, the lack of application of existing knowledge about nutrients and soil management continues to be a major problem in the developing countries where most of the growth in fertilizer use must occur. There is a tremendous need for both the scientific and nonscientific communities to realize that blanket recommendations for fertilizers (i.e., recommendations for an area rather than individual fields) cause inefficient use of fertilizers, and are detrimental to soils and the environment. As the most yield-limiting macronutrients are applied and yields are increased, previously nonlimiting nutrients will become deficient, and it will become very important to be able to determine the deficiencies through soil and plant tissue testing. Excessive build-up of soil fertility has not only occurred in developed countries, but also in countries like Indonesia. While this problem can be avoided with soil testing, many other countries are apparently not applying enough P and K to maintain soil fertility. Since soil testing has not been used to monitor fertility changes, these trends are not known, and it is difficult to determine the reason for yield stagnation or decline, such has been occurring with irrigated rice in Southeast Asia.

CONCLUSIONS

The world population has increased from less than 2 billion people at the turn of the century to 5.7 billion in 1995, and it is expected to reach 8.5 billion in 2025. This unprecedented growth in population will create tremendous pressures on the natural resource base to produce enough food and fiber to meet human needs and wants. The application of improved crop production technologies has resulted in a global food supply which kept ahead of food demand during the 1960-90 period. Many of the food-deficit countries, particularly in Asia, greatly improved their ability to produce more food stuffs during a period when food prices have been relatively low. Fertilizers have played a critical role in this achievement, and will be critical in meeting future food requirements. Maintaining the soil resources of Africa poses particularly difficult challenges. Lack of incentives, infrastructure problems, and rapid currency devaluations have greatly discouraged fertilizer use.

Adequate food supplies in the future will require food grain production to increase from current levels of 1,950 million tonnes to 3,100-3,500 million tonnes by 2020. In addition, production of root and tuber crops, fruit, vegetables, oil seed, and fiber crops must also increase by 40-45%. There is limited scope for expanding the cultivated or irrigated area in most regions of the world, so the additional food and fiber production must come from increased crop yields on the existing arable land. In raising crop yields, the use of fertilizers will be indispensable. Global fertilizer production and use will need to increase from 123 million nutrient tonnes to over 300 million nutrient tonnes in 2020 to meet food needs. At projected growth rates, fertilizer use is expected to reach only 208 million nutrient tonnes.

The demand for more food production, particularly in the most disadvantaged areas, mandates that we apply existing agronomic and policy knowledge more effectively to promote fertilizer use and greater efficiency of use. There are still many improvements in fertilizer use, production, marketing, and fertilizer-related policy that could greatly contribute to food security, alleviation of poverty, and resource conservation in developing countries. Policy initiatives that try to address root causes for the soil fertility decline and soil degradation are urgently needed in sub-Saharan Africa.

REFERENCES

Baanante, C.A., B.L. Bumb, and T.P. Thompson. (1989). *The Benefits of Fertilizer Use in Developing Countries.* Paper Series No. P-18, IFDC, Muscle Shoals, Alabama, USA.

Blackie, M.J. (1995). Maize productivity for the 21st century: the African challenge. In *Proc. Fourth Eastern and Southern Africa Regional Maize Conference,* eds. D.C. Jewell, S.R.Waddington, J.K. Ranson, and K.V. Pixley. Harare, Zimbabwe, 28 March-1 April, 1994. Mexico City, Mexico: CIMMYT, pp. 11-22.

Bongaarts, J. (1994). *Global and Regional Population Projections to 2025.* Paper presented at the IFPRI Round Table on Population and Food in the Early 21st Century, Washington, D.C., USA, February 14-16.

Borlaug, N.E. and C.R. Dowswell. (1993). Fertilizer: To nourish infertile soil that feeds a fertile population that crowds a fragile world. *Fertilizer News* 387: 11-20.

Breman, H. (1995). *Opportunities and Constraints for Sustainable Development in Semi-Arid Africa.* The United Nations University, Institute for New Technologies (UNU/INTECH) Working Paper No.18, Maastricht, The Netherlands.

Brown, L.R. and H. Kane. (1994). *Full House.* New York, USA: W.W. Norton.

Bumb, B.L. (1994). *Global Fertilizer Perspective, 1980-2000: The Challenges in Structural Transformation.* Muscle Shoals, Alabama, USA: International Fertilizer Development Center.

Bumb, B.L. and C.A. Baanante. (1996a). *The Role of Fertilizer in Sustaining Food Security and Protecting the Environment to 2020.* Food, Agriculture, and the Environment Discussion Paper 17. Washington, D.C., USA: International Food Policy Research Institute.

Bumb, B.L. and C.A. Baanante. (1996b). *World Trends in Fertilizer Use and Projections to 2020.* Brief 38, Paper 1. Washington, D.C., USA: International Food Policy and Research Institute.

Easterbrook, G. (1997). Forgotten benefactor of humanity. *The Atlantic Monthly,* January, 74-82.

Engelstad, O.P. and D.T. Hellums. (1993). *Water Solubility of Phosphate Fertilizers: Agronomic Aspects–A Literature Review.* Paper 17. Muscle Shoals, Alabama, USA: International Fertilizer Development Center.

Food and Agriculture Organization (FAO). (1996a). *19th Regional Conference for Africa. World Food Summit: Policy Statement and Plan of Action.* Rome, Italy: FAO.

Food and Agriculture Organization (FAO). (1996b). Production data diskette. Rome, Italy: FAO.

Gerner, H. and G. Harris. (1993). The use and supply of fertilizers in sub-Saharan Africa. In *The Role of Plant Nutrients for Sustainable Food Crop Production in Sub-Saharan Africa,* eds. H. Van Reuler and W.H. Prins. Leidschendam, The Netherlands: VKP, pp. 107-125.

Hammond, L.L., S.H. Chien, and A.U. Mokwunye. (1986). Agronomic value of unacidulated and partially acidulated phosphate rock indigenous to the tropics. *Advances in Agronomy* 40: 89-140.

Harris, J.M. (1996). World agricultural futures: Regional sustainability and ecological limits. *Ecological Economics* 17: 95-115.

Hazell, P.B.R. (1994). *Prospects for a Well-Fed World.* Washington, D.C., USA: International Food Policy Research Institute (Draft).

Hitzhusen, F.J. (1993). Land degradation and sustainability of agricultural growth: Some economic concepts and evidence from selected developing countries. *Agriculture, Ecosystems and Environment* 46: 69-79.

International Fertilizer Association (IFA). (1995). Fertilizer data diskette. Paris, France: IFA.

International Fertilizer Development Center (IFDC). (1996). *Restoring and Maintaining the Productivity of West African Soils: Key to Sustainable Development.* Miscellaneous Fertilizer Studies No. 14. Muscle Shoals, Alabama, USA: International Fertilizer Development Center.

International Food Policy Research Institute (IFPRI). (1995). *A 2020 Vision for Food, Agriculture, and the Environment: The Vision, Challenge, and Recommended Action.* Washington, D.C., USA: International Food Policy Research Institute.

International Fund for Agricultural Development (IFAD). (1993). *The State of World Rural Poverty, a Profile of Africa.* Rome, Italy: International Fund for Agricultural Development.

Larson, B. (1993). *Fertilizers to Support Agricultural Development in Sub-Saharan Africa: What Is Needed and Why.* Center for Economic Policy Studies Discussion Paper No. 13. Washington, D.C., USA: Winrock International.

Malthus, T.R. (1798). *Essays on the Principle of Population.* London, UK: Oxford University Press.

Mellor, J.W. (1988). The intertwining of environmental problems and poverty. *Environment* 309: 8-13, 28, 29.

Paddock, W. and P. Paddock. (1967). *Famine, 1975.* Boston, Massachusetts, USA: Little, Brown and Co.

Pinstrup-Andersen, P. (1993). Future perspectives on food supply in developing countries. *Outlook on Agriculture* 224: 225-232.

Pinstrup-Andersen, P. and R. Pandya-Lorch. (1994). *Alleviating Poverty, Intensifying Agriculture, and Effectively Managing Natural Resources.* Food, Agriculture and the Environment Discussion. Paper 1. Washington, D.C., USA: International Food Policy Research Institute.

Ponting, C. (1990). Historical perspectives on sustainable development. *Environment* 329: 4-9, 31-33.

Population Reference Bureau (PRB). (1995). World Population Data Sheet. Washington, D.C., USA: Population Reference Bureau.

Pouzet, D. (1991). Profitability of fertilizer use options in Ghana. In *Fertilizer Use at the Village Level: Constraints and Impacts.* Summary Proceedings of Workshop. Lomé, Togo: International Fertilizer Development Center, pp. 45-47.

Quinones, M.A., N.E. Borlaug, and C.R. Dowswell. (1996). Fertilizer-based green revolution for Africa. Paper presented at the American Agronomy Society Meeting, Indianapolis, Indiana, USA, November 3-8.

Rosegrant, M.W., M. Agcaoili-Sombilla, and N.D. Perez. (1995). *Global Food Projections to 2020: Implications for Investment.* Food, Agriculture, and the Environment Discussion Paper 5, Washington, D.C., USA: International Food Policy Research Institute.

Schultz, J.J., D.I. Gregory, and O.P. Engelstad. (1993). *Phosphate Fertilizers and the Environment–A Discussion Paper*. Paper 16. Muscle Shoals, Alabama, USA: International Fertilizer Development Center.

Seckler, D. (1994). *Trends in World Food Needs: Toward Zero Growth in the 21st Century*. Center for Economic Policy Studies Discussion Paper No. 18. Arlington, Virginia, USA: Winrock International Institute for Agricultural Development.

Smaling, E.M.A. (1993). Soil nutrient depletion in sub-Saharan Africa. In *The Role of Plant Nutrients for Sustainable Food Crop Production in Sub-Saharan Africa*, eds. H. Van Reuler and W.H. Prins. Leidschendam, The Netherlands: VKP.

Smaling, E.M.A. (1995). The balance may look fine when there is nothing to mine: Nutrient stocks and flows in West African soils. In *Use of Phosphate Rock for Sustainable Agriculture in West Africa*, eds. H. Gerner and A.U. Mokwunye. Muscle Shoals, Alabama, USA: International Fertilizer Development Center, pp. 10-20.

Speth, J.G. (1993). *Towards Sustainable Food Security*. Sir John Crawford Memorial Lecture to the Consultative Group on International Agricultural Research, 25 October 1993.

Stoorvogel, J.J. and E.M.A. Smaling. (1990). *Assessment of Soil Nutrient Depletion in Sub-Saharan Africa, 1983-2000*. Report 28, Wageningen, The Netherlands: Winand Staring Centre for Integrated Land, Soil and Water Research.

United Nations (UN). (1992). *Long-Range World Population Projections: Two Centuries of Population Growth*, 1950-2150. New York, USA: United Nations.

Vaidyanathan, A. (1993). *Fertilizers in Indian agriculture*. L.S. Venketaraman Memorial Lecture, Institute for Social and Economic Change, Bangalore, India.

Van der Pol, F. (1993). Analysis and evaluation of options for sustainable agriculture, with special reference to southern Mali. In *The Role of Plant Nutrients for Sustainable Food Crop Production in Sub-Saharan Africa*, eds. H. van Reuler and W.H. Prins. Leidschendam, The Netherlands: VKP, pp. 69-88.

Van Kauwenbergh, S.J. (1995). Overview of the global phosphate rock production situation. In *Direct Application of Phosphate Rock and Appropriate Technology Fertilizers in Asia–What Hinders Acceptance and Growth*, eds. S.J. Van Kauwenbergh and D.T. Hellums. Proc. Workshop IFS and IFDC, Kandy, Sri Lanka, Feb. 20-25, pp. 15-28.

Van Reuler, H. (1996). *Nutrient Management Over Extended Cropping Periods in the Shifting Cultivation System of South-West C^ote d' Ivoire*. Ph. D. Thesis, Agricultural University, Wageningen, The Netherlands.

Vlek, P.L.G. (1993). Strategies for sustaining agriculture in sub-Saharan Africa: The fertilizer technology issue. In *Technologies for Sustainable Agriculture in the Tropics*. ASA Special Publication 56, eds. J. Ragland and R. Lal. Madison, Wisconsin, USA: American Society of Agronomy, Crop Science Society of America, Soil Science Society of America, pp. 265-277.

World Bank. (1992). *World Development Report, 1992: Development and the Environment*. Washington, D.C., USA: World Bank.

World Commission on Environment and Development (WCED). (1987). *Our Common Future.* Geneva, Switzerland: World Commission on Environment and Development.

Zaman, M.S. (1991). Famine–causes and health consequences. In *The Challenge of African Disasters*, New York, USA: WHO and UNITAR, United Nations, pp. 77-107.

SUBMITTED: 02/13/97
ACCEPTED: 06/10/97

Managing Soil Fertility Decline

Lindsay C. Campbell

SUMMARY. Soil fertility is defined in terms of the ability of the soil to maximize plant productivity, often within economic constraints. A decline in natural soil fertility seems to have occurred over all civilizations. Overgrazing and deforestation are the two most important factors affecting global soil degradation. Nutrient losses from agricultural systems are broadly divided into losses through volatilization, losses from leaching, losses due to product removal, losses to non-labile soil pools and losses from various forms of erosion. Rapid declines in soil fertility are associated with large demands for food due to expanding population, nutrient mining of agricultural areas with concomitant shifts of produce to cities, and intensification of agricultural activities without proper regard for long-term maintenance of fertility by application of fertilizers, recycling of organic wastes, liming to combat acidification, fallowing, rotations and prevention of large scale soil erosion. Agricultural policy has often encouraged soil fertility decline and soil degradation. In the future, agricultural scientists must have major inputs into the development and implementation of policy. *[Article copies available for a fee from The Haworth Document Delivery Service: 1-800-342-9678. E-mail address: getinfo@haworthpressinc.com]*

KEYWORDS. Agricultural policy, land degradation, nutrient budget, nutrient loss, soil erosion, soil fertility, sustainability, yield loss

WHAT IS SOIL FERTILITY?

Soil fertility is an implicitly defined concept. Although the literature refers to soil fertility, definitions are largely lacking. Generally, soil

Lindsay C. Campbell, Associate Dean, Faculty of Agriculture, University of Sydney, Sydney NSW 2006, Australia (E-mail: lindsay.campbell@cropsci.su.edu.au).

[Haworth co-indexing entry note]: "Managing Soil Fertility Decline." Campbell, Lindsay, C. Co-published simultaneously in *Journal of Crop Production* (Food Products Press, an imprint of The Haworth Press, Inc.) Vol. 1, No. 2 (#2), 1998, pp. 29-52; and: *Nutrient Use in Crop Production* (ed: Zdenko Rengel) Food Products Press, an imprint of The Haworth Press, Inc., 1998, pp. 29-52. Single or multiple copies of this article are available for a fee from The Haworth Document Delivery Service [1-800-342-9678, 9:00 a.m. - 5:00 p.m. (EST). E-mail address: getinfo@haworthpressinc.com].

29

fertility has connotations of plant productivity often coupled with economic overtones. In the last couple of decades, soil fertility has often been equated with "soil health" or subsumed into the topic of "soil health." DeLuca and Jacobsen (1988) at Montana State University defined soil fertility as: "Soil's ability to supply essential nutrients in adequate amounts and proportions for plant growth throughout growing season." A common defacto measure of soil fertility is (crop) yield.

When plants are obtaining their optimum (mineral) nutrition, it is usually then that pest and disease resistance is greatest. Nutrient availability and water availability are perhaps the two most important factors influencing plant growth. A typical curve showing yield responses to added nutrients is presented in Figure 1. Yields without added nutrient supply as a fertilizer, amendment or organic material such as manure or compost give an estimate of the inherent or basic soil fertility. Curves can be fitted to the yield response function; the most commonly fitted curves are the Mitscherlich equation or a quadratic function or a combination of both (Bock and Sikora, 1990). The Mitscherlich equation is given by

$$y = a(1 - be^{-cF})$$

where y is the predicted yield, a is the maximum likely yield in a given situation, b is the soil responsiveness co-efficient having a domain between 0 and 1, c is the curvature co-efficient and F the rate of fertilizer applied. The Mitscherlich equation does not account for toxicity, whereas a quadratic function handles toxicity. However, the latter is often not the best description of the zones of deficiency.

If the Mitscherlich curve is projected through to zero yield, then the value obtained between zero and this intercept is a measure of the available nutrient in the soil without any added nutrient. Highly fertile soils have a large (negative) intercept. A positive intercept can be interpreted as strong tie-up or binding of the nutrient to the soil colloids, matrix or soil organic matter. It is important for plant nutritionists, agronomists and horticulturalists to generate response curves at least to the zone of luxury consumption. The shape of the yield response curve is generally similar from year to year but the actual maximum yields may vary considerably. A detailed yield response curve can then be linked to economic production functions where the

FIGURE 1. Soil fertility as a function of the native amount of nutrients in the soil and applied nutrients such as fertilizers, amendments (e.g., lime) or organic matter.

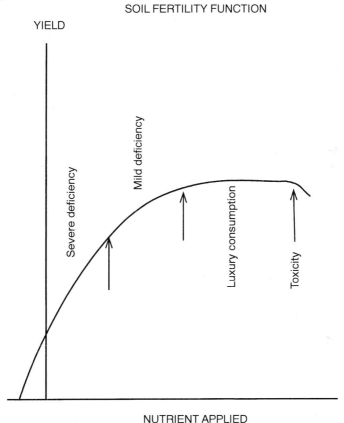

cost of inputs and the value of the commodity produced may fluctuate widely within and between years.

AGRICULTURAL POLICY

Agricultural and other governmental policies can directly affect changes in soil fertility due to influences on the behaviour of farmers or indirectly influence the marketing and distribution chain or consumers. The debate between public and private benefits is likely to affect agricultural policy greatly in the 21st century (cf. Just and Rausser, 1993). In essence, issues such as–who captures the benefits of agricultural research, who profits from land care and soil erosion programs, who benefits from improved human health due to better foods–will dominate policy decisions and thus affect local decisions concerning the productivity, fertility and maintenance of the land. Although it can be shown that farmers may profit directly from a new plant protection agent, or a company benefit financially from the release of a new cultivar, the community also stands to benefit from higher quality produce or comparatively low food prices. Low food prices often tend to give greater political stability and thus greater economic productivity for the whole country.

Inappropriate policy decisions can often have adverse consequences. During the First World War, the Australian rural workforce declined by some 27%, making for less labor intensive farming (Davidson, 1981). After that war, the Australian government opened up a considerable areas of land to soldiers returning from the war. Unfortunately, allotments were generally too small for broad-acre cropping and often the incoming discharged soldiers were not trained for agriculture and were undercapitalised (Davidson, 1981). As well as the social misery generated (Lake, 1987), land degradation occurred through poor practice and the necessity of making a living from an inadequate acreage. Following the Second World War, a similar scheme was instituted but land grants were larger and generally there were capital requirements placed on the applicants as well as appropriate experience. Commentators rate this scheme as more successful than the first (Campbell, 1982) but part of this success may be related to the high prices of commodities after the war.

Perhaps on account of the land areas allocated, dairy farms were often the favoured enterprise. Many coastal dairy farms were established but were marginal economically. Although subsidies were introduced, land care was not a high priority. As these farmers age and dairying subsidies are removed, land amalgamation to large economically-viable allotments is taking place. With larger, economically-viable units, investment in the property has taken place, and there are definite signs of improvement in soil fertility.

In the Australian context, land ownership by the farmer is often perceived as the best way to minimize land degradation. The reasoning is that greater care of one's own property occurs and that the farm is left in good condition for the children's inheritance. On the other hand, land renting and share farming are often practised in the US, allowing farmers to find off-farm income without necessarily involving a shift in the population from rural to urban areas. However, the percentage of the population involved in agricultural production is now much less than 5% in both Australia and the US (Campbell, 1994).

An interesting land care (and indirectly soil fertility) policy is pursued in France: farmers are to be stewards of the land, and hence such social objectives of maintenance of the land may have ramifications in other areas of policy, e.g., subsidies.

Hess and Holechek (1995) have examined the influence of policy on land degradation in the arid western region of the US. In essence, government acts of the 19th and early 20th century limited the legal size of holdings of ranchers to units which were not sustainable. Overgrazing occurred on both public and private lands with concomitant loss of soil through erosion. The grazing permits were introduced to control overgrazing, but these permits formalized excessive numbers of livestock, perhaps on account of political influence on the allocations. Permits are valuable as they can be traded, unlike the land. Is the solution to overgrazing to reduce the livestock-carrying capacity attached to a permit or to increase the cost of permit to uneconomic levels so that grazing is no longer profitable, or for the public purse to buy out the ranchers? Hess and Holechek also note that alternative land uses may provide a better cash flow and be more sustainable, but the current legislation mitigates against the alternatives.

Policy decisions often change farmers' decision making. A bounty on superphosphate will encourage its use: the consequences may range from moving from native to improved pasture, growing crops with high phosphate requirements, use of marginal land which should not be in production, to excessive applications of fertilizer beyond any productive use, or to increased run-off of soil rich in bound phosphates, which in turn pollute waterways. A policy, commenced as an incentive to improve soil phosphorus fertility and thus crop and pasture productivity, may have resulted in both positive and negative effects. Bounties or subsidies on nitrogenous fertilizers are even more likely to have adverse environmental repercussions.

Droughts not only influence farmers' decisions but often result in land degradation through exposure of cultivated bare land for long periods and overgrazing, both of which lead to soil erosion. A classic example of this is the erosional losses of soil in 1930's in huge dust storms in the western Nebraska and neighbouring states. Plant diversity may also change, especially in grazing lands: palatable grasses are preferentially grazed and seed production is reduced. Woody weeds tend to increase as evidenced in the arid regions of the western US, Africa and Australia. Policies of drought relief to farmers have often encouraged farmers and ranchers at the margin to remain in production rather than having these lands returned to other uses, including recreational uses.

Population shifts from rural areas to cities have many implications for loss of (agriculturally) productive land. For example, the huge expansion of the Egyptian population in the Nile Delta has resulted in considerable areas of fertile land no longer being used for agricultural production. Likewise, in the Okanagan region of British Columbia, one of the most agriculturally important areas of western Canada, retirees are moving in large numbers into this productive area with resulting loss of quality land to housing or small holdings. Policies need to be developed quickly to maintain highly productive agricultural lands. However, a policy which says that certain land is only for agricultural activity is particularly difficult to implement, especially when the value of agricultural activities is far less that the value of the land for housing, hobby farms or recreation.

Population shifts are mainly from inland rural to urban areas, as evidenced in Australia with most people living in cities on the coastal fringe. Although not as extensive, depopulating the mid-western states and provinces is occurring in the US and Canada. Iowa, in the highly productive mid-west of the US, is unusual as it is going against this trend of depopulating inland agricultural states. Large population centres can be viewed as nutrient sinks because food, rich in nutrients, is transported to the cities for human consumption, while subsequent recycling of these nutrients is limited. Policies are needed to deal with this nutrient shift; in some cases, treated municipal effluent can be recycled for agricultural purposes.

Sites of agricultural industries can mean major shifts in long-term fertility and affect the sustainability of the practices. Therefore, what may be needed is the development of regenerative food systems (Dahlberg, 1993); it may ultimately mean local production as well as consumption. Intensive animal industries can potentially lead to major changes in soil fertility as evidenced in Japan (Harada, 1992), Europe and the US. For example, dairying in Pennsylvania and poultry raising in Georgia result in large quantities of grain being imported into those states and thus large inputs of nutrients. Export of dairy or poultry products does not offset the nutrient inputs, i.e., a net accumulation of nutrients is occurring. Firstly, there is the issue of loss of nutrients from mainly mid-western states in the US in what may be termed "nutrient mining" of the soils; secondly, there is the issue of how to dispose of, or recycle, the nutrients in the net importing states. Manure can be spread over some parts of the farm to increase fertility, but generally the amounts of manure produced are in excess of the plant productivity of the farm. This can lead to pollution of waterways and ground waters, especially with nitrogen. Is the solution to transport nutrients away to less fertile areas, or should production be more widely distributed over the country, or should the industry be located close to the grain source? Economics have largely driven the decision of location; embedded in the economic decision are factors relating to climate, e.g., raising poultry in a relatively mild climate, where supplementary heating or cooling for much of the year is not required, reduces costs of production. The availability of cheap land may dictate the site without consideration of the soils at that location. What has not been costed are changes in soil fertility, environmental pollution, or the cost of nutrient removal to other localities or states.

Commodity prices can drastically alter farming systems and soil fertility. Farms which have rotations including legumes in the rotation will abandon all rotations in years of high wheat prices. This affects many parameters in the soil including soil nitrogen, soil microorganisms, organic matter and cation exchange capacity, especially on lighter soil types. Commodity prices are often driven by consumers, e.g., fashions in wool or cotton have resulted in rapid price increases with resultant changes in local farming systems. There is no easy solution to large price signal changes.

The Aswan Dam in Egypt has given many benefits in terms of irrigation potential, but the downside has been the prevention of flood waters carrying fertile soils to the delta, the deposition of those sediments, and increases in soil salinity resulting in desertification. Wolff (1986) has argued that the nutrients carried in depositional soil in the Nile Delta were insufficient to meet the total nutrient requirements of more intensive agricultural production and thus fertilizer inputs would be required. Notwithstanding Wolff's conclusion, irrigation water, although limited, has raised the water table and soil salinity; drainage of saline waters is necessary to improve soil fertility, yet these waters are often reused for agriculture (Moustafa et al., 1987; El-Mowelhi et al., 1988; Abdel-Dayem and Abu-Zeid, 1992; Abbott et al., 1996). Some of these problems may have been avoided if scientists and policy makers each had a knowledge of the other discipline.

With a rapidly increasing global population, food security policies should be established for both developed and less developed countries, irrespective of whether the country is a net agricultural exporter or importer. Most developing countries have food security policies in place. The role of fertilizers should be addressed in those policies (Byrnes and Bumb, 1998). Soil fertility and its maintenance should feature strongly in such plans. Given current agricultural policies of many developed countries, it is likely that major shortfalls of food will occur within the next 15-20 years.

NUTRIENT LOSSES

Nutrient losses from agricultural systems can be broadly divided into losses through volatilization, losses from leaching, losses due to crop (or product) removal, losses to non-labile soil pools and losses from various forms of erosion. Volatilization losses are chiefly due to burning (of stubble), gaseous injection of ammonia fertilizers or microbial degradation leading to volatile products. Leaching losses are dependent on rainfall and its intensity, amount of irrigation, soil water permeability, and plant rooting depth.

Volatilization Losses

Burning of stubble is widely practised around the world either for reducing the biomass of the crop residue, controlling weeds or minimizing disease. Due to atmospheric pollution, stubble burning is being phased out in some states, e.g., burning rice stubble in California. The practice of stubble burning is used to control *Pyrenophora* (yellow leaf spot) in wheat in Australia (Clarke and Gagen, 1988; Martin, McMillan, and Cook, 1988), and stem rot (*Sclerotium oryzae*) in rice in California (Dobie, Miller, and Mosley, 1984), and for limited control of downy brome (Rasmussen, 1995). Suppression of Pythium and Rhizoctonia root rots occurred when stubble was burnt, provided soil fertility was high (Smiley, Collins, and Rasmussen, 1996). During combustion of the stubble, nutrients–particularly nitrogen and sulfur– are lost to the atmosphere (Table 1); on the other hand, most of the minerals are retained either on the ground or in airborne particles which usually do not drift more than 10 to 100 km. After fire, the pH of the soil–particularly the top 1-5 cm increases by more than 1 pH unit (Rundel and Parsons, 1980) due to the hydrolysis of the oxides of potassium and calcium. In very hot fires, phosphorus can react with the soil minerals (Attiwill and Leeper, 1990) and thus become unavailable for plant growth. Nutrients lying virtually on the soil surface after fire are prone to be washed away if heavy rains occur before recolonisation takes place.

In slow-growing forests, nutrient losses through fire may be significant in themselves, but the long-term losses may not be that great due to replenishment from parent soil rocks and minerals, atmospheric deposition (especially of nitrogen and phosphorus to a small extent), and nitrogen fixation by legumes or other nitrogen-fixing species. However, regular burning of stubble or prescribed burning of forests, as occurs in parts of California and eastern Australia, may result in decreases of soil fertility and decreases in the soil microflora and microfauna. Fertilizers and/or legumes can assist in rebuilding this fertility.

Estimates of the nitrogenous losses due to volatilization from soils are most imprecise because of a high degree of variability. Mosier,

TABLE 1. Estimates of nutrient losses from burning straw residues, or from forest fires

Crop	Biomass (t/ha)	Nitrogen (kg/ha)	Sulfur (kg/ha)
Rice straw (California)	8	65-120	4.5-8
Wheat straw (NSW)	5	75-130	7-18
Pine forest (Canada)*	200	1,000	200
Eucalyptus forest (Australia)*	400	500-2,000	100-300

*Biomass of forests vary due to age, climate, species, cultivation, etc. The pine forest biomass is taken from Prescott, Corbin, and Parkinson (1989). A range of eucalyptus biomasses is given in Birk and Turner (1992). The biomass of large forests in the US has reached over 850 t/ha (Satoo and Madgwick, 1982). Nutrient loss values represent complete combustion and therefore are overestimations for most forest fires.

Chapman, and Freney (1989) added 80 kg N/ha as ^{15}N-enriched urea to flooded rice at the panicle initiation stage. They found daily flux densities of nitrous oxide, ammonia and dinitrogen to be less than 5, 170 and 720 g N/ha respectively, i.e., a relatively small loss of about 1.2% occurred per day. In this situation, about 95% of the labelled urea could be recovered either as gases, within the plant, in the soil, or in the water. On the other hand, considerable nitrogenous losses–perhaps as much as half the applied urea–occurred in similar flooded rice soils experiencing a series of flooding and drying cycles (Bacon et al., 1986). Nitrate leaching was low; losses were attributed to denitrification and to incorporation into soil organic matter from previous crops.

Nitrous oxide formation is more likely to be greater with urea than nitrate as the fertilizer source (Duxbury, 1990). Denitrification accounted for about 85% of the total nitrogen lost from nitrate fertilizers (Duxbury and McConnaughey, 1986). On the other hand, a considerable amount of nitrous oxide was formed during nitrification of ammonium ions liberated from urea. Overall, the net loss of nitrogen was greater from urea than nitrate (Duxbury and McConnaughey, 1986).

Recovery of nitrogenous fertilizers by crops is often about 50% (Obi, Hedlin, and Cho, 1986). In a series of experiments, Adjetey (1991) found that nitrogen recoveries by wheat crops growing on a red brown earth soil (Natrixeralf) were about 50%, even when conditions existed for substantial carry-overs of nitrogen from one season to the next. Urea was the source of applied nitrogen fertilizer; placement depths varied but the gains from deep placement were marginal and certainly not economic. Similar recoveries are reported for a vertisol (Saffigna et al., 1984). Adjetey (1991) also showed that there was little downwards transport of nitrogen (measuring both nitrate and ammonium) beyond 30 cm in the profile, yet roots occurred to a depth of 90 cm.

Volatilization of other nutrients is likely to be insignificant. Cations and most anions of agricultural importance are not volatile; in theory, reduction of sulfate can occur, but reduced sulfur compounds are likely to react with soil water or soil minerals in most situations. Occasionally, under anaerobic conditions, hydrogen sulphide gas is released.

Losses from Product Removal

One of the most important losses of nutrients with a consequent depletion of soil fertility is transportation of plant products away from the site of production. Usually, this does not result in localised nutrient cycling. Using Australian wheat as an example of an adverse change in fertility, both yields and protein concentration in the grain have declined (Dalal et al., 1991; Verrell and O'Brien, 1996) with considerable economic consequences. To a large extent, this is due to lower levels of available nitrogen in the soils. Corrective measures are increased use of fertilizer nitrogen or rotations with legumes. The latter may have the added benefit of disease control or improving soil structure; however, it is difficult to quantify these benefits in simple, monetary terms.

Grain is usually transported long distances from the site of production to points of processing or ultimate consumption. When exporting grain overseas, it is interesting to note the cost paid by the purchaser includes a portion for nutrients. It may amount to US $20 per tonne of grain for high quality wheats. Nutrient removal in some agricultural products is given in Table 2. The amounts of macronutrients removed in grain, pastures and fruit can be quite significant, especially when considered over a number of years. No soil can remain productive unless such removal of nutrients is replenished, often with inorganic

TABLE 2. Nutrient removal by crops, pastures, fruits and livestock products

Commodity	Wheat	Alfalfa	Banana	Livestock	Milk
Yield	2.5 (t/ha)	2.5 (t/ha per cut)	45 (t/ha)	per 1,000 kg live-weight**	Per cow/ year*
Nutrient					
N (kg/ha)	63	95	45	16	16
P (kg/ha)	8	10	10	25	8
K (kg/ha)	10	88	100	5	10
S (kg/ha)	4	4	10	4	2
Mg (kg/ha)	3	11	11	1	0.8
Ca (kg/ha)	1	50	50	38	9
Zn (g/ha)	96	100	50	125	10
Cu (g/ha)	9	25	25	10	1
Fe (g/ha)	120	175	175	150	3
Mn (g/ha)	116	125	125	1	1
B (g/ha)	6	88	88	–	1

*No dry period is assumed. Trace element concentrations vary widely in milk (*vide* Jenness, 1985).
**Sulfur is about 4% in wool, hence sulfur values should be higher for sheep than cattle.

fertilizers. Hamblin and Kyneur (1993) have documented the decline in wheat yields in Australia between 1870 and 1900: yields declined by 57% during this period because of no inputs of phosphorus or nitrogen. When superphosphate was applied widely, yields recovered to the initial cleared-land levels of productivity. Application of nitrogen, or rotations with legumes, coupled with better weed control, gave a further improvement in yield of more than 30%.

Rates of mineralisation of nutrients from parent materials are far too low to sustain any macronutrient removal on the scale indicated in Table 2. Initial plant growth is achieved by the bank of soil nutrients. This bank is largely a function of organic matter, soil cation exchange capacity, and depth, as well as the parent materials. On some vertisols, phosphorus requirements of crops such as low-yielding wheats in continuous annual production have been met for half a century or more without fertilizer addition. However, when more intensive farming is practised, including double cropping, yields decline after 10 to 15 years and nutrient deficiency symptoms appear. This is because the readily available soil nutrients are being taken up faster than the rate of release of nutrients from other less labile pools. Fallowing is one way of partially restoring soil fertility: organic matter is broken down and nutrients released to more available forms, accompanied by a small release of nutrients from parent rocks and minerals. It is interesting to note that, 4000 years ago, the proscription was to fallow every seventh year and fallow for two years in the 49th and 50th years (Bible, 1989).

In many countries, shifting cultivation and "slash and burn" practices are simply an expression of a decline in soil fertility. Slashing and burning release nutrients from the trees, shrubs and grasses. Some of the nutrients are washed away on the bare soil and much of the remainder is taken up by subsequent plant growth. The net result is a less fertile soil which may take many decades to return to its former level of fertility.

For the same weight of product removed, the quantities of the cations potassium, magnesium and calcium are three to fifty times more for alfalfa than wheat, whereas the quantities of the anions phosphate and sulfate removed are similar in both species (calculated from the data of Reuter and Robinson, 1986; Jones, Wolf, and Mills, 1991; Wiese, 1993). Large demands for cations are common in the legumes; also roots of legumes have a much higher cation exchange capacity than grasses. The net result of a dissimilar removal of cations to anions is a change in soil pH. When excess cations are removed, soil acidification occurs.

Nutrient losses from horticultural crops are often quite high due to high yields on an area basis and moderately high concentrations of nutrients in the products. When production is near to villages, e.g., in parts of Indonesia, there is the potential for much of the nutrients to be recycled locally. On the other hand, horticultural industries in developed countries are usually located long distances from consumers and thus the opportunities for nutrient recycling are usually low. Extensive exporting of horticultural produce is common in several South American countries. This commodity export results in nutrient mining, depletion of soil fertility and increased soil acidity unless appropriate measures are taken to offset the nutrient removal. Remedies may include fertilizer inputs matching removal, liming (including the use of dolomite to maintain the calcium-to-magnesium ratio in the soil), green manure crops and mulches. Usually, the cost of remediation measures is not factored into the cost of production, and thus long-term sustainability is compromised. In my view it is better to provide incentives to implement remediation measures than to raise the price of the foods produced. The latter tend to affect the poor and malnourished more than other sectors of a population, and high food prices are often political and social destabilisers.

Removal of nutrients in livestock or animal products is not insignificant. However, a more important nutrient removal problem is the transport of nutrients either from one part of the farm to another or the import of grain or hay to feed animals in more intensive situations. The former is well illustrated in dairies with the paddocks near the milking sheds having much higher fertility than more distant paddocks due to the proportionally greater amounts of animal manure excreted. Similarly, sheep camps are highly fertile with respect to nutrients, but are usually degraded with respect to vegetation and soil compaction.

Estimates of nutrient requirements can be made using two different approaches. One is to estimate the requirements based on ion uptake theory, the simplest of which is to invoke the Nernst Equation or Flux Equations which are based on electrochemical potential differences driving the uptake of ions. A variant to this electrochemical approach is Bouldin's ion uptake model (Bouldin, 1989), which could be developed further. Notwithstanding its limitations, this ion uptake model often fits well with observed values (Bouldin, Miyasaka, and Grunes,

1992). Generally, ion uptake models are much better predictors of cation uptake than anion uptake, but Bouldin's model does attempt to deal with this problem by stressing the role of dissolved CO_2. This difference in predicting cation and anion uptake is because of the electrical potential difference across the membrane–the inside is negatively charged with respect to the outside. Thus, cations migrate towards the negative charge, whereas anions are repelled.

An alternative approach is to calculate nutrient requirements based on, for example, grain yield, fruit yield, or dry matter production coupled with a knowledge of nutrient concentrations in these fractions. Although total nutrient requirements can be estimated for above-ground plant parts, there is a lack of good data on root/shoot ratios during development and nutrient concentrations in the roots.

Losses to Non-Labile Pools

Some soils, especially those with high amounts of iron or aluminium clay minerals, have a capacity to "fix" phosphate. Phosphate-fixing soils often occur in the tropics, South America, Africa and Australia. Many of these soils are also characterised by a low pH. Chemical reactions occur to form iron or aluminium phosphates. These reactions to form insoluble phosphates are energetically favourable, with large changes in the Gibbs free energy. The iron phosphate is more stable than the aluminium phosphate (Friesen and Blair, 1988). As a consequence of the very stable phosphate compounds, the release of phosphate is an extremely slow process. Management of these soils to minimise fixation of phosphate can include liming to reduce the free aluminium or hydrated aluminium ions present, growing species more tolerant of low soil phosphate concentrations (e.g., lupins can access phosphate pools unavailable to soybeans, Braum and Helmke, 1995), and timing the application of phosphate fertilizer to match the demands of the plants. The form of phosphate fertilizer affects immediate uptake, thus readily soluble forms may be used in times of high demand. On the other hand, slow-release rock phosphate may be beneficial in releasing phosphate to slow-growing pastures. Broadcasting phosphate is less efficient for crops than banding.

Phosphate fixation can also occur on alkaline soils, particularly those having high levels of calcium; the result can be the formation of calcium phosphate. There has been debate as to whether nitrification can increase the availability of fixed phosphate. Faurie and Fardeau

(1990) concluded that nitrification and its associated acidification were of no importance under high demands for phosphate, such as in the case of cropping or improved pastures.

Losses to relatively non-labile pools can occur with the slow decomposition of organic matter, e.g., nitrogen and sulphur can enter pools which are broken down over a period of years. Leaching of a nutrient or even deep placement beyond the rooting zone can also be considered in this category.

LAND DEGRADATION AND SOIL EROSION

Land degradation in all its various forms is not something new. Desertification processes such as erosion and salinity in the Middle East and North Africa have been recorded from ancient civilizations. Dailey (1995) estimates that 43% of vegetated land is degraded; surprisingly, degradation is not a function of the economic wealth of a country, occurring in developed and less developed countries alike. Dailey argues that overgrazing and deforestation account for over 65% of the land degradation; a further 28% is due to other forms of agriculture. Land degradation by animals is a source of concern in the US, especially in the more arid areas (see example in Hess and Holechek, 1995). In Australia, the introduction of the rabbits caused much of the land degradation as they ate extensive tracts of native and introduced pastures. In addition, the development of watering holes for livestock also meant that kangaroos then had ready access to water, with a concomitant explosion in their numbers. Desertification of Australia's arid and semiarid lands now accounts for some 5×10^6 km^2 (Ludwig and Tongway, 1995), noting that much of this area has long been in this category.

Desertification is one of the major processes of land degradation globally. Its occurrence is widespread from Africa and the Middle East to China, the Indian subcontinent, South America, Australia and the US. It impacts most on those countries that experience periodic severe droughts. Livestock eating the sparse vegetation during drought contribute substantially to the land degradation: the removal of vegetation results in little seed set, loss of species diversification, and instability of the soil, thus making it susceptible to erosion. Fuel-wood collection is another important contributor to increasing desertification. Usually with desertification, there is a concomitant loss of soil fertility. Solutions to the problem, such as reducing stocking rates, fencing, planting for fuel-wood, etc., are often not readily acceptable socio-economically. For example, in Argentina, beef production gains occurred under reduced grazing pressure but there were substantial capital outlays (Guevara and Estevez, 1995), which may not be achievable for farms most at risk of declining soil fertility. Governments may have to invest heavily in control or income support measures if a long-term goal of recovery is to be achieved.

Soil erosion is usually the most visible sign of land degradation and reduced soil fertility. It often shows as gullies, rills or blown soil accumulations or dust storms, and is therefore more likely to receive press coverage. Losses of soil of about 10-20 t/ha translate into substantially less than 1 mm of surface soil lost, which is still greater than the gains in soil from mineralisation of parent materials. At these commonly accepted rates of erosion, even if extrapolated over decades, it is difficult to rationalise the soil losses with the attributed losses in yield. It would seem that there are properties of the surface soil about which little is known; this would seem to be a profitable area for future research.

Soil erosion and loss of surface nutrients are minimised by having good ground cover (cf. Unger and Agassi, 1996), including mulches (Edwards, Burney, and DeHaan, 1995). Maize grown in no-till situations has comparatively low rates of soil erosion (McDowell and McGregor, 1984; Lindstrom, 1986). Lindstrom notes that a form of nutrient mining occurred when soil loss levels were in excess of 11 t/ha/year coupled with nutrient losses from removal of the crop, i.e., net fertilization rates were less than the combined losses due to soil loss and crop removal. Figure 2 shows that in tilled soil, erosional losses are directly correlated to runoff. On the other hand, there is no correlation with run-off for no-till soil.

The practice of intercropping, widely used in many tropical countries, needs to be evaluated for temperate regions prone to surface runoff of nutrients and soil. Intercropping red clover with maize in Ontario, Canada, showed promise both in terms of erosion control and the benefits of fixed nitrogen from the clover (Wall, Pringle, and Sheard, 1991). However, benefits from use of legumes intercropped with other species, e.g., bananas, may reduce yields excessively (Johns, 1994).

Soil fertility can decline due to many other factors, e.g., acid soils,

FIGURE 2. Erosion from runoff of water from tilled soil. Derived from data of McDowell and McGregor (1984).

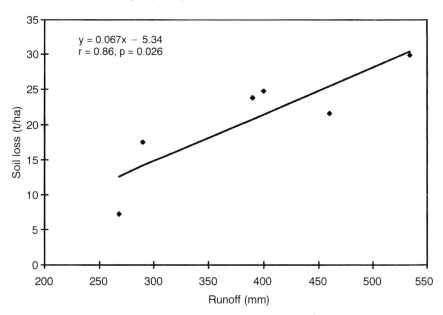

acid sulfate soils, salinity, etc. (Since much is written on salinity, it is not covered here except to note that increasing salinity is often a function of poor water management, land clearing (especially on hill tops), application of waters high in dissolved salts, and rising water tables. Trees are often considered as a means of lowering the water tables.) Increases in soil acidity can be attributed to acid rain (an important factor in Europe, parts of North America and areas near large industrial centers), product removal such that excess cations to anions are removed, nitrate leaching from the rooting zone, application of acidifying fertilizers such as ammonium sulfate or urea, increasing load of organic matter that is not being recycled, application of acidic irrigation water, and leaching of cations in already acidic soils. There is no good evidence to show that applications of superphosphate acidify the soil. In fact, McLaughlin and James (1991) and Sloan, Basta, and Westerman (1995) have used banded phosphate as a possible means of alleviating aluminium toxicity in acid soils.

Soil acidity problems tend to be greater in the tropics, South America (especially Brazil), and parts of eastern Australia than in many of the cool temperate climates. In Europe, the practice of liming is well established from Roman times.

Removal of products rich in cations relative to anions is probably one of the greatest sources of long-term acidification. Increasingly, transport of products away from the source of production is occurring without recycling of the nutrients to the source location. Thus, it is expected that more acid soil problems will occur in the future, unless liming occurs. Slattery, Ridley, and Windsor (1991) suggest that a lime equivalent for a wheat crop yielding 2 t/ha amounts to 19 kg lime/ha. To maintain an equilibrium soil pH in subterranean clover pastures, about 40 kg/ha of lime would have to be applied annually, whereas several hundred kg/ha of lime would have to be applied to alfalfa crops cut for hay and transported away from the site of production (Ridley, Heylar, and Slattery, 1990; Slattery, Ridley, and Windsor, 1991).

The increase in soil acidity under legumes is due to product removal (either grain or forage), accumulated organic matter or to breakdown of nitrogen-rich organic matter to nitrate which is subsequently leached beyond the rooting zone. Growing deep rooted perennials has been suggested as a way of minimizing nitrate leaching (Ridley and Windsor, 1992). Noble, Zenneck, and Randall (1996) have noted that recycling nutrients from depth is at least a nominal way of bringing alkalinity to the soil surface. At least some tree species can contribute to net reduction in acidity.

Acid sulfate soils have their own unique problems and occur world-wide. It is only comparatively recently that this topic has received much attention. Oxidation of pyrite, often associated with drainage of a soil, results in the production of sulfuric acid. Usually under these conditions there is an increase in toxic aluminium ions, and sometimes high manganese and iron; phosphorus is usually low, as is calcium. Thus, it is not surprising that plants are particularly sensitive to acid sulphate soils. Liming may be seen as one solution to the problem by both increasing the pH and by providing a source of calcium which assists in decreasing the root-cell membrane permeability and improving the selectivity for the uptake of ions.

THE FUTURE

Nitrification is a topic that is still not well understood, and the dynamics of nitrogen transformations in the soil, although described, is one that can benefit from further research. One thrust is to have ammonium- and nitrate-nitrogen fertilizers stored into, say, a moderately labile organic pool. This may reduce nitrification losses and leaching of nitrate to ground waters.

As nitrogen is required in large amounts by plants, the use of legumes to provide a source of nitrogen is likely to increase dramatically. Legumes are not the only source of fixed nitrogen; it is possible that the problems of nitrogen fixation in wheat may be overcome within a decade. Other microorganisms (Frankia, blue green algae) are capable of fixing nitrogen in symbiotic relationships. There is some evidence indicating that sugarcane may obtain considerable amounts of its nitrogen from associations with microorganisms. Careful management of any plant or agricultural system fixing nitrogen will need to occur to avoid changes in soil pH.

Much more research needs to be done on various forms of soilless agriculture, including hydroponics. Although this promises to hold good prospects for intensive production, e.g., vegetable and floricultural production, considerable work needs to be done on reutilizing spent water rich in some nutrients, removal of waste products and disease minimization in recycled nutrient solutions. Likewise, the role of soil microflora and microfauna should be much further researched; for example, earthworms, belonging to many genera, are very important in recycling nutrients, yet have not been widely appreciated by agricultural scientists.

Nutrient budgets will need to be prepared for farms, catchments, states and countries. Through this process, the nutrient transport chains can be determined and possibilities for recycling of nutrients explored. It may mean that agriculture and urbanization may have to be more closely linked. It is likely that the whole question of sustainability will have to be re-examined. Nutrient mining must have a finite lifetime. Can a system which moves nutrients with only some replenishment continue? Furthermore, energy costs associated with urbanization and transport of nutrients may rise sufficiently to rethink the rural-to-urban population shift.

Soil acidity, including acid sulfate soils, is a continuing problem. Part of the solution may be in the judicious use of lime or dolomite, irrigation management to minimise differential leaching of nutrients, and developing more acid-tolerant species. The last approach is likely only to have effects at the margin. Providing conditions whereby a net increase in hydroxyl anions occurs during decomposition of organic matter may be another approach. When nutrients are recycled *in situ*, such as litter decomposition in forests, the net changes in soil acidity are minimal, provided there are no losses of nutrients from the system. Can more nutrient cycling be promoted in relatively intensive agricultural production with its large shifts of nutrients to distant markets? Given the technological advances in communication, is it becoming feasible for the population of a city to be more geographically dispersed to minimise nutrient shifts?

Although it is not specifically research activity, scientists need training in policy, including agricultural policy. Likewise, policy makers must have a grasp of scientific and technological issues related to soil fertility and agriculture in general. A variant on such training is the implementation of specialist teams. However, a team must be able to communicate clearly and team members must understand the background language of the discipline areas of the other members and a grasp of the fundamentals of the discipline. It is my belief that the greatest gains in improvement of global and local soil fertility during the next decade will occur if scientists and policy makers are grounded in one another's disciplines. Reward systems are required to encourage individuals to span diverse discipline areas. Currently, there are no rewards and often there are disincentives. What a challenge to universities and other institutions!

REFERENCES

Abbott, C.L., D.E.D. El-Quosy, G.R. Pearce, and M.N. Bayoumi. (1996). Soil salinity levels due to drainwater reuse in the Nile Delta. Proceedings of 6th Drainage Workshop on Drainage and the Environment, Ljubljana, Slovenia, April 21-29, 1996. New Delhi, India: International Commission on Irrigation and Drainage, pp. 523-533.

Abdel-Dayem, S. and M. Abu-Zeid. (1992). Salt load in irrigation and drainage water in the Nile Delta. Proceedings of the African Regional Conference on Technologies for Environmentally Sound Water Resources Development, Alexandria, Egypt, 1991. Vol. 2. Wallingford, UK: Wallingford Ltd., pp. 131-142.

Adjetey, J.A. (1991). *Effects of Nitrogen on Yield and Grain Quality of Wheat (Triticum aestivum L.) Grown on a Red-Brown Earth Soil.* Ph.D. Thesis. The University of Sydney, Sydney, Australia.

Attiwill, P.M. and G.W. Leeper. (1990). *Forest Soils and Nutrient Cycles.* Melbourne, Australia: Melbourne University Press.

Bacon, P.E., J.W. McGarity, E.H. Hoult, and D. Alter. (1986). Soil mineral nitrogen concentration within cycles of flood irrigation: effect of rice stubble and fertilization management. *Soil Biology and Biochemistry* 18: 173-178.

Bible. (1989). Leviticus 25 in *Holy Bible*. Iowa Falls, Iowa, USA: World Bible Publishers.

Birk, E.M. and J. Turner. (1992). Response of flooded gum (*E. grandis*) to intensive cultural treatments: biomass and nutrient content of eucalypt plantations and native forests. *Forest Ecology and Management* 47: 1-28.

Bock, B.R. and F.J. Sikora. (1990). Modified-quadratic/plateau model for describing plant responses to fertilizer. *Soil Science Society of America Journal* 54: 1784-1789.

Bouldin, D.R. (1989). A multiple ion uptake model. *Journal of Soil Science* 40: 309-319.

Bouldin, D.R., S.C. Miyasaka, and D.L. Grunes. (1992). Cation accumulation by winter wheat forage: II. Correlation with multi-ion model. *Journal of Plant Nutrition* 15: 1081-1097.

Braum, S.M. and P.A. Helmke. (1995). White lupin utilizes soil phosphorus that is unavailable to soybean. *Plant and Soil* 176: 95-100.

Byrnes, B.H. and B.L. Bumb. (1998). Population growth, food production and nutrient requirements. In *Nutrient Use in Crop Production*, ed. Z. Rengel. Binghamton, New York, USA: The Haworth Press, Inc., pp. 1-27

Campbell, K.O. (1982). Land policy. In *Agriculture in the Australian Economy*, ed. D.B. Williams. 2nd ed. Sydney, Australia: Sydney University Press, pp. 225-239.

Campbell, L.C. (1994). Beneficial impact of precision nutrient management on the environment and future needs. *Communications in Soil Science and Plant Analysis* 25: 889-908.

Clarke, R.G. and S. Gagen. (1988). Incidence of yellow leaf spot (*Pyrenophora tritici repentis*) of wheat in Victoria. *Plant Protection Quarterly* 3: 16-18.

Dahlberg, K.A. (1993). Regenerative Food Systems: broadening the scope and agenda of sustainability. In *Food for the Future: Conditions and Contradictions of Sustainability*, ed. P. Allen. New York, USA: John Wiley & Sons, pp. 75-102.

Dailey, G.C. (1995). Restoring value to the world's degraded lands. *Science* 269: 350-354.

Dalal, R.C., W.M. Strong, E.J. Weston, and J. Gaffney. (1991). Sustaining multiple production systems. 2. Soil fertility decline and restoration of cropping lands in sub-tropical Queensland. *Tropical Grasslands* 25: 173-180.

Davidson, B.R. (1981). *European Farming in Australia*. Amsterdam, The Netherlands: Elsevier.

DeLuca, T.H. and J.S. Jacobsen. (1988). Glossary of soil fertility terms. A-2 Fertilization. Montguide MT8803. Bozeman, Montana, USA: Montana State University Extension Service.

Dobie, J.B., G.E. Miller and R.H. Mosley. (1984). Ground level harvest of rice straw. *Transactions of the American Society of Agricultural Engineers* 27: 1263-1269.

Duxbury, J.M. (1990). Agriculture, nitrous oxide, and our environment. *New York's Food and Life Sciences Quarterly* 20: 28-31.

Duxbury, J.M. and P.K. McConnaughey. (1986). Effect of fertilizer source on denitrification and nitrous oxide emissions in a maize field. *Soil Science Society of America Journal* 50: 644-648.

Edwards, L., J. Burney, and R. DeHaan. (1995). Researching the effects of mulching on cool-period soil erosion control in Prince Edward Island, Canada. *Journal of Soil and Water Conservation* 50: 184-187.

El-Mowelhi, N., A. El-Bershamgy, G.J. Hoffman, and A.C. Chang. (1988). Enhancement of crop yields from subsurface drains with various envelopes. *Agricultural Water Management* 15: 131-140.

Faurie, G. and J.C. Fardeau. (1990). Can acidification associated with nitrification increase available soil phosphate or reduce the rate of phosphate fixation? *Biology and Fertility of Soils* 10: 145-151.

Friesen, D.K. and G.J. Blair. (1988). A dual radiotracer study of transformations of organic, inorganic and plant residue phosphorus in soil in the presence and absence of plants. *Australian Journal of Soil Research* 26: 355-366.

Guevara, J.C. and O.R. Estevez. (1995). Cattle industry: a sustainable development alternative for the arid zones of Argentina. *Decertification Control Bulletin* 27: 58-61.

Hamblin, A. and G. Kyneur. (1993). *Trends in Wheat Yields and Soil Fertility in Australia*. Canberra, Australia: Australian Government Printing Service.

Harada, Y. (1992). Composting and land application of animal wastes. *Asian Australasian Journal of Animal Sciences* 5: 113-121.

Hess, K. and J.L. Holechek. (1995). Policy roots of land degradation in the arid region of the United States: an overview. *Environmental Monitoring and Assessment* 37: 123-141.

Johns, G.G. (1994). Effect of *Arachis pintoi* groundcover on performance of bananas in northern New South Wales. *Australian Journal of Experimental Agriculture* 34: 1197-1204.

Jenness, R. (1985). Biochemical and nutritional aspects of milk and colostrum. In *Lactation*, ed. B.L. Larson. Ames, Iowa, USA: Iowa State University Press, pp. 164-197.

Jones, J.B., B. Wolf, and H.A. Mills. (1991). *Plant Analysis Handbook*. Athens, Georgia, USA: Micro-Macro Publishing.

Just, R.E. and G.C. Rausser. (1993). The governance structure of agricultural science and agricultural economics. *American Journal of Agricultural Economics* 75: 69-83.

Lake, M. (1987). *The Limits of Hope*. Melbourne, Australia: Oxford University Press.

Lindstrom, M.J. (1986). Effects of residue harvesting on water runoff, soil erosion and nutrient loss. *Agriculture, Ecosystems and Environment* 16: 103-112.

Ludwig, J.A. and D.J. Tongway. (1995). Desertification in Australia: an eye to grass roots and landscapes. *Environmental Monitoring and Assessment* 37: 231-237.

Martin, R.J., M.G. McMillan, and J.B. Cook. (1988). Survey of farm management practices of the northern wheat belt of New South Wales. *Australian Journal of Experimental Agriculture* 28: 499-509.

McDowell, L.L. and K.C. McGregor. (1984). Plant nutrient losses in runoff from conservation tillage corn. *Soil and Tillage Research* 4: 79-91.

McLaughlin, M.J. and T.R. James. (1991). Effects of surface-applied phosphorus and superphosphate on the solution chemistry and phytotoxicity of subsurface aluminium: sand/solution and soil experiments. *Australian Journal of Agricultural Research* 42: 859-874.

Mosier, A.R., S.L. Chapman, and J.R. Freney. (1989). Determination of dinitrogen emission and retention in floodwater and porewater of a lowland rice field fertilized with ^{15}N-urea. *Fertilizer Research* 19: 127-136.

Moustafa, A.T.A., W.H. Ahmed, A.C. Chang, and G.J. Hoffman. (1987). Influence of depth and spacing of tile drains on crop productivity in the Nile Delta. *Transactions of the American Society of Agricultural Engineers* 30: 1374-1377.

Noble, A.D., I. Zenneck, and P.J. Randall. (1996). Leaf litter ash alkalinity and neutralisation of soil acidity. *Plant and Soil* 179: 293-302.

Obi, A.O., R.A. Hedlin, and C.M. Cho. (1986). Crop utilization of nitrogen from ^{15}N-labelled urea, calcium nitrate and ammonium sulphate in several Manitoba soils. *Canadian Journal of Soil Science* 66: 661-671.

Prescott, C.E., J.P. Corbin, and D. Parkinson. (1989). Biomass, productivity, and nutrient-use efficiency of aboveground vegetation in four Rocky Mountain coniferous forests. *Canadian Journal of Forest Research* 19: 309-317.

Rasmussen, P.E. (1995). Effects of fertilizer and stubble burning on downy brome competition in winter wheat. *Communications in Soil Science and Plant Analysis* 26: 951-960.

Ridley, A.M., K.R. Helyar, and W.J. Slattery. (1990). Soil acidification under subterranean clover (*Trifolium subterraneum* L.) pastures in north-eastern Victoria. *Australian Journal of Experimental Agriculture* 30: 195-201.

Ridley, A.M. and S.M. Windsor. (1992). Persistence and tolerance to soil acidity of phalaris and cocksfoot in north-eastern Victoria. *Australian Journal of Experimental Agriculture* 32: 1069-1075.

Reuter, D.J. and J.B. Robinson. (1986). *Plant Analysis: An Interpretation Manual.* Sydney, Australia: Inkata Press.

Rundel, P.W. and D.J. Parsons. (1980). Nutrient changes in two chaparral shrubs along a fire induced age gradient. *American Journal of Botany* 67: 51-58.

Saffigna, P.G., A.L. Cogle, W.M. Strong, and S.A. Waring. (1984). The effect of stubble on ^{15}N fertilizer nitrogen balance and immobilization in a vertisol during the summer fallow. In *The Properties and Utilization of Cracking Clay Soils*, eds. J.W. McGarity, E.H. Hoult, and H.B. So. Armidale, Australia: University of New England, pp. 218-220.

Satoo, T. and H.A.I. Madgwick. (1982). *Forest Biomass.* The Hague, The Netherlands: Martinus Nijhoff.

Slattery, W.J., A.M. Ridley, and S.M. Windsor. (1991). Ash alkalinity of animal and plant products. *Australian Journal of Experimental Agriculture* 31: 321-324.

Sloan, J.J., N.T. Basta, and R.L. Westerman. (1995). Aluminium transformations and solution equilibria induced by banded phosphorus fertilizer in acid soil. *Soil Science Society of America Journal* 59: 357-364.

Smiley, R.W., H.P. Collins, and P.E. Rasmussen. (1996). Diseases of wheat in long-term agronomic experiments at Pendleton, Oregon. *Plant Disease* 80: 813-820.

Unger, P.W. and M. Agassi. (1996). Common soil and water conservation practices. In *Soil Erosion, Conservation and Rehabilitation*, ed. M. Agassi. New York, USA: Marcel Dekker, pp. 239-266.

Verrell, A.G. and L. O'Brien. (1996). Wheat protein trends in northern and central NSW, 1958 to 1993. *Australian Journal of Agricultural Research* 47: 335-354.

Wall, G.J., E.A. Pringle, and R.W. Sheard. (1991). Intercropping red clover with silage corn for soil erosion control. *Canadian Journal of Soil Science* 71: 137-145.

Wiese, M.V. (1993). Wheat and other small grains. In *Nutrient Deficiencies and Toxicities in Crop Plants*. ed. W.F. Bennett. St. Paul, Minnesota, USA: APS Press, pp. 27-33.

Wolff, P. (1986). Der Nilschlamm und sein Einfluss auf die Fruchtbarkeit der Ackerboden in Agypten. *Tropenlandwirt* 87: 143-161.

SUBMITTED: 03/27/97
ACCEPTED: 07/08/97

Soil and Plant Testing Programs as a Tool for Optimizing Fertilizer Strategies

Peter J. Van Erp
Marinus L. Van Beusichem

SUMMARY. Soil and plant testing programs are still based on 'trial and error' methods and lack scientific underpinning in terms of relevant soil chemical and plant nutritional processes, and are site-specific. The programs are valuable when the objective is to diagnose and predict deficiencies of plant nutrients. The programs are less valuable for refined fertilizer strategies, like Integrated Nutrient Management, which are essential in the near future to satisfy changing agricultural, environmental, economic and legislative boundary conditions. A more scientific approach to soil and plant testing programs appears desirable. To reduce undesirable side-effects of fertilization on the environment, more emphasis should be placed on fertilizer type and on timing and method of application. *[Article copies available for a fee from The Haworth Document Delivery Service: 1-800-342-9678. E-mail address: getinfo@haworthpressinc.com]*

KEYWORDS. Crop nutrition, fertilizer recommendations, plant analysis, plant testing programs, soil analysis, soil testing programs

Peter J. Van Erp, Senior Soil Scientist, Nutrient Management Institute, Wageningen Agricultural University, Wageningen, The Netherlands. Marinus L. Van Beusichem, Associate Professor of Plant Nutrition, Department of Soil Science and Plant Nutrition, Wageningen Agricultural University, Wageningen, The Netherlands.

Address correspondence to: Marinus L. Van Beusichem, Department of Soil Science and Plant Nutrition, Wageningen Agricultural University, P.O. Box 8005, NL 6700 EC Wageningen, The Netherlands (E-mail: Rien.vanBeusichem@bodvru.benp. wau.nl).

[Haworth co-indexing entry note]: "Soil and Plant Testing Programs as a Tool for Optimizing Fertilizer Strategies." Van Erp, Peter J., and Marinus L. Van Beusichem. Co-published simultaneously in *Journal of Crop Production* (Food Products Press, an imprint of The Haworth Press, Inc.) Vol. 1, No. 2 (#2), 1998, pp. 53-80; and: *Nutrient Use in Crop Production* (ed: Zdenko Rengel) Food Products Press, an imprint of The Haworth Press, Inc., 1998, pp. 53-80. Single or multiple copies of this article are available for a fee from The Haworth Document Delivery Service [1-800-342-9678, 9:00 a.m. - 5:00 p.m. (EST). E-mail address: getinfo@haworthpressinc.com].

GENERAL INTRODUCTION

In order to feed the growing world population, agricultural crop production has to increase considerably. To attain this, efforts should be focussed on increasing crop yields per hectare rather than increasing the area for agricultural production (World Bank, 1992; IFA, 1995). Improvement of the fertility status of agricultural land and an economic, efficient (re-)use of mineral and organic fertilizers, organic wastes and crop residues should be promoted to achieve increases in crop yields (Smaling, 1993; Van Reuler, 1996).

Agriculture in European and North-American regions is characterized by a high crop production, a (more than) sufficient soil fertility status and a high input of nutrients via mineral and organic fertilizers (IFA, IFDC, and FAO, 1994; FAO, 1995). In these regions, agriculture is confronted with the (in)direct side effects of current management leading to nutrient losses to the environment (Isermann, 1990; Bussink, 1992; 1994), adverse effects on product quality (Breimer, 1982), high energy inputs (Fluck, 1992), production of greenhouse gases (Granli and Bøckman, 1994; Koops, Oenema, and Van Beusichem, 1996; Velthof, Brader, and Oenema, 1996), acidification (Oenema, 1990), etc.

Faced with these side effects, The Netherlands (Anonymous, 1987) and the EC (EC, 1991) proposed legislation that restricts rate, time and method of nutrient applications and the nitrogen (N) and phosphorus (P) surplus on the N and P balance sheet of farms (VROM and LNV, 1995). Its aim is to optimize nutrient-use efficiency and minimize negative side effects. To achieve compliance with an increasing amount of agricultural, environmental, legislative and economic constraints, there is a need for well-defined fertilizer strategies (Van Erp and Oenema, 1993). These strategies should lead to optimization of nutrient use, crop production and quality and at the same time satisfy the above-mentioned boundary conditions. Fertilizer strategies can be based on: (1) soil testing programs that relate nutrient availability in the rooting zone, in space and time, to crop demand (De Willigen and Van Noordwijk, 1987; Slangen, Titulaer, and Rijkers, 1989); and (2) plant testing programs that monitor crop nutrient content during growth, allowing corrective fertilizer application (Munson and Nelson, 1990).

In this chapter, the design and scientific underpinning of current soil and plant testing programs will be discussed for macronutrients and annual field crops. The perspectives in using such programs as practical tools for optimizing fertilizer strategies will then be evaluated.

COMPONENTS OF SOIL AND PLANT TESTING PROGRAMS

There is a common agreement that 17 chemical elements are essential for metabolism, growth, development and successful reproduction of higher plants (Epstein, 1965, 1972; Brown et al., 1987; Marschner, 1995). Insight into the dynamics of nutrient availability in soil and crop nutrient requirement is necessary to optimize soil fertility status, to synchronize supply and demand and thus to maximize crop yield. Soil and plant testing programs can be useful practical tools in reaching these goals.

Soil and plant testing programs include: (1) collection and preparation of soil and plant samples; (2) chemical extraction (or pressing) of the samples; (3) determination of the nutrient concentration in the extract; (4) interpretation of the obtained nutrient concentrations in order to assess soil fertility categories or plant status categories; and (5) derivation of (corrective) fertilizer applications (Dahnke and Olson, 1990; Munson and Nelson, 1990; Peck and Soltanpour, 1990).

We define soil and plant analysis as the chemical/physical treatment of the soil or plant sample and subsequent determination of the nutrient concentration. Soil and plant analysis data provide the basis for the fertilizer recommendations, and thus form an essential part of soil and plant testing programs.

Background to Soil and Plant Analysis

Soil Analysis

Schofield (1955) distinguished two nutrient fractions in the soil: the 'quantity,' indicating the amount of potentially available nutrients, and the 'intensity,' indicating the strength of nutrient retention. The 'quantity' reflects all the nutrients within or adsorbed at the soil constituents, whilst the 'intensity' reflects the nutrient concentration in the soil solution. The 'intensity' and 'quantity' are interrelated by the buffering capacity of the soil, which is an indicator of the capability to maintain a certain nutrient concentration in solution. The 'quantity'/'intensity' approach is valuable for nutrients like P and K (Holford, 1991; Holford and Doyle, 1992; Evangelou, Wang, and Phillips, 1994; Raven and Hossner, 1994), but cannot easily be applied to nutrients predominantly in organic forms and/or to nutrients that are hardly buffered by soil constituents. The concentration of (non-buffered) nutrients in the soil solution may vary enormously because of fertilization, nutrient uptake by crops and mineralization (Figure 1; see also Yanai et al., 1996).

Nutrient uptake rate by plant roots is considered to be positively correlated with the nutrient concentration in the soil solution (Nye and Tinker, 1977; Barber, 1984), i.e., with the 'intensity.' The nutrient

FIGURE 1. Dynamics of the concentration of Ca, NO_3, Mg, K and P as well as pH in the soil solution (Yanai et al., 1996, with kind permission from Kluwer Academic Publishers, Dordrechet, The Netherlands). ○: without N without plant, ●: without N with plant, △: with N without plant, ▲: with N with plant.

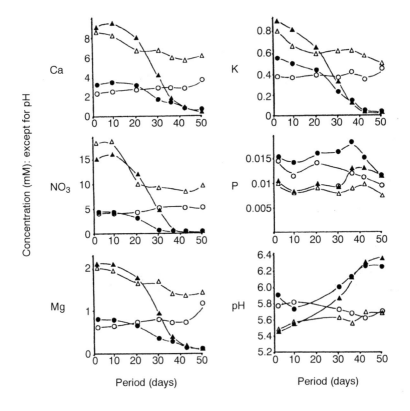

concentration in soil solution may thus be a good indicator of the actual nutrient availability in the soil. The methods available to separate the soil solution from soil constituents (Dahlgren, 1993; Jones and Edwards, 1993; Lorenz, Hamon, and McGrath, 1994; Lawrence and David, 1996) do not always provide actual concentrations because the soil solution may be altered substantially during the separation process. Nevertheless, soil extraction with water or dilute salts (Houba et al., 1990; Dahlgren, 1993) is widely used to assess the nutrient concentration in soil solution. When using these weak extractants, the amounts of extracted nutrients heavily depend on e.g., sample drying temperature (Figure 2) and sample storage (Barlett and James, 1980; Houba, Novozamsky, and Van der Lee, 1989, 1995; Rechcigl, Payne, and Sanchez, 1992), soil:solution ratio and shaking time (Rezaian et al., 1992), and extraction temperature (Houba, Novozamsky, and Van der Lee, 1989). Results of soil extraction with water or dilute salt solutions are probably related, but certainly not equal to the actual nutrient concentration in the soil solution. Interpretation/quality of soil testing programs may improve if the soil chemical processes that determine the nutrient release during the extraction process are taken into account.

Determination of the 'quantity' can be done by means of total elemental analysis. From a crop nutritional point of view, the significance of these total analyses is limited because only a very small fraction of the total reserve can be taken up by the crop during one growing period. From an agricultural point of view, estimation of the size of the 'labile' (Marschner, 1995) pool may be a better indicator of the nutrient availability. Extractants commonly used to determine this 'labile' pool are (combinations) of acids, hydroxides, complexing agents or salt solutions (Fixen and Grove, 1990; Haby, Russelle, and Skogley, 1990). Also ion-exchange resins (Rubaek and Sibbesen, 1993) or ion-exchange membranes (Qian, Schoenau, and Huang, 1992) are sometimes used to determine the size of the 'labile' nutrient pool. The theoretical foundation of the functioning of most extractants is well known, but it is difficult to use this knowledge for selecting an extractant because the chemical binding forms of nutrients in the soil are mostly unknown. Generally, nutrients associated with the cation exchange complex are extracted with high molar salt solutions (Haby, Russell, and Skogley, 1990; Meyer and Arp, 1994). Nutrients that are present in minerals with a low solubility product, or in minerals from

FIGURE 2. Effect of drying temperature on the amount of NO_3-N, NH_4-N and soluble organic N extracted with 0.01 M $CaCl_2$ (Houba, Novozamsky, and Van der Lee, 1989, with kind permission from VDLUFA, Darmstadt, Germany).

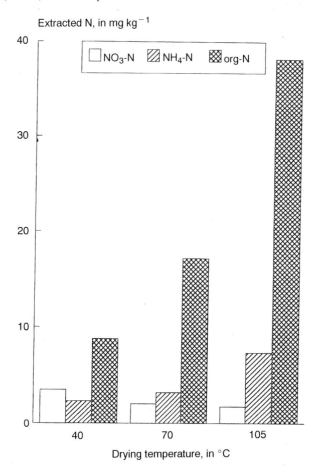

which the release is kinetically restricted, are extracted using acids or hydroxides, resins or other nutrient-specific methods (Fixen and Grove, 1990; Menon, Chien, and Chardon, 1997).

Plant Analysis

Plant nutrients are mostly taken up from the soil in an ionic form (Mengel and Kirkby, 1987). After uptake, the nutrients are distributed throughout the plant but the major part is transported towards growing cells with an active metabolism (Marschner, 1995). Determination of the nutrient content of the whole plant is generally not a good indicator of its nutritional status because a substantial proportion of the nutrient is not metabolically active and/or incorporated in cell structures. An expanded but not fully mature leaf is metabolically very active and, therefore, it is considered that its nutritional composition may be used in the diagnosis of the nutritional state of most crops (Martin-Prével, Gagnard, and Gautier, 1987; Jones and Case, 1990; Marschner, 1995).

Advances in analytical techniques, procedures and equipment, and the increased knowledge of physiological plant nutrition have extended the development of foliar analysis as a basis of plant testing programs. In most cases, the total elemental content of the leaves is analyzed in oven-dried, ground plant material. However, extraction of plant samples with water, solutions of acetic acids, dilute HCl or a mixture of HF and HCl are also in use (Jones and Case, 1990).

Tissue testing may involve determination of the nutrient concentration in plant sap squeezed from fresh plant samples. In this way semi-quantitative information can be obtained about plant nutrients such as NO_3, HPO_4/H_2PO_4, and K (Jones and Case, 1990).

The presence or activity of enzymes or nutrient-containing metabolites may be related to the plant nutritional status (Bar-Akiva, 1971; 1984; Bouma, 1983; Hernandez et al., 1995). However, enzyme activity is not always nutrient-specific and may also be affected by plant age and external factors (Bar-Akiva, 1971, 1984; McLachlan, 1982). In addition, these techniques are labour intensive and, therefore, their usefulness in plant testing programs is limited in the short term.

Development of Soil and Plant Testing Programs

The development of soil and plant testing programs can be divided into five phases (Dahnke and Olson, 1990; Munson and Nelson, 1990): (1) formulation of soil and plant sampling strategies; (2) assessment of the correlation between the amount of nutrient extracted and crop yield or nutrient uptake; (3) ranking into soil fertility or plant status categories (= calibration); (4) interpretation of results of pot and field trials and recommendations of fertilizer rates, and (5) adjustment of the fertilizer recommendations to economic boundary conditions.

Phase 5 is an integration of phase 4 with financial boundary conditions and is beyond the scope of this work.

Sampling Strategy

A bulked soil sample needs to reflect/represent the spatial heterogeneity of the soil in an agricultural field in both horizontal and vertical directions. Numerous soil sampling procedures have been proposed for obtaining a representative soil sample from spatially heterogeneous fields (James and Wells, 1990). Similary, a plant sample needs to reflect the heterogeneity of the performance of the crop in the field and the nutritional status of a crop. Crop-specific sampling procedures have been proposed (Jones and Case, 1990). Most plant sampling procedures have an empirical basis.

Correlation Studies

Determining the best soil or plant extractant, traditionally, relies on the determination of the relationship between the concentration of the nutrient extracted and crop yield or nutrient uptake (Corey, 1987). Extractants fail when the nutrient concentration is not, or is only weakly, related to crop yield or nutrient uptake. Correlation research is usually conducted in two steps: exploratory (fertilizer) trials in the greenhouse followed by trials in the field. The advantage of pot experiments in the greenhouse is that the possible effects of the conditions in the subsoil, weather conditions and soil heterogeneity on crop yield or nutrient uptake can largely be eliminated. When an extractant is successful in greenhouse experiments, the relationship needs to be tested further in field trials because crop response is a function of many variables (Dahnke and Olson, 1990). Data of field experiments may vary because of the many factors that determine yield, e.g., soil, crop, weather, management, etc. Therefore, results of correlation analysis are often improved when relative, rather than absolute, crop yield or nutrient uptake is plotted (Dahnke and Olson, 1990; Holford and Doyle, 1992).

The perspectives of a newly proposed soil or plant extractant can be assessed by correlating the extraction results obtained with the new extractant to those obtained using the standard extractant (Houba et al., 1990; Matejovic and Durackova, 1994). Although this is a very useful first-step technique in evaluating a new extractant, results need to be interpreted cautiously.

Soil Fertility Categories and Plant Status Categories

To simplify the process of making fertilizer recommendations, the results of soil and plant analysis are ranked in categories (Dahnke and Olson, 1990). A common procedure is to plot results of soil or plant analysis vs. (relative) crop response and to fit a continuous curve through the points. The curve can then be divided into soil fertility categories such as very low, low, medium, high and very high (Hauser, 1973), and plant status categories such as severe deficiency, mild deficiency, luxury range and toxic range (Figure 3; see also Smith, 1962). The basis for the division is mostly subjective and arbitrary. A more objective alternative to establish soil fertility categories is the probability approach (Fitts, 1955). This approach builds on the assumption that the results of soil analysis are not more than an indication of the crop response probability. The graphical Cate-Nelson method (Cate and Nelson, 1965) separates soils that respond from those that do not respond to added nutrients. This method has been presented as a statistical procedure that can be used to establish two or more categories (Cate and Nelson, 1971; Nelson and Anderson, 1977).

FIGURE 3. Relationship between nutrient concentration in the leaves and the growth or yield of the crop, and the division of nutrient concentrations into plant status categories (Adapted from Smith, 1962).

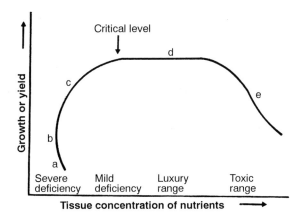

Interpretation of Analyses and Development of Fertilizer Recommendations

Soil Testing

In interpreting soil analytical results obtained from pot and field experiments, the relationship between the amount of nutrient extracted, the nutrient application rate and crop yield should be established. Such relationships are normally described by response models such as the linear model concept of Liebig (Waggoner and Norvell, 1979) and the curvilinear model of Mitscherlich (1928). Generally, curvilinear models are more suitable for the interpretation of field data and development of fertilization recommendations. These models, based on the 'Law of Diminishing Returns,' imply that when equal increments of a nutrient are applied to a crop, the yield response becomes smaller for each increment. This type of crop response is found in many field and pot trials. The relationship between fertilizer application and crop yield will normally be improved when soils are grouped in soil fertility categories as established by soil test calibration (Hauser, 1973). In this way, each soil fertility category has its own curve that relates nutrient application rates to crop responses.

A general criticism of curvilinear models for interpretation of field data and development of fertilizer recommendations is that in the region of near maximum yield they recommend too much fertilizer in relation to the possible increase in crop yield. In the 'Plateau yield' method (Dahnke and Olson, 1990), the relationship between fertilizer application rate and crop yield is assessed according to the linear-model concept (Waggoner and Norvell, 1979). The linear-model concept shows more clearly the application rate at which maximum yield (the plateau) is reached. From an agricultural point of view, it is logical to apply nutrients corresponding to this yield plateau.

During the 1940s and 1950s, the cation saturation ratio concept was proposed (Bear and Toth, 1948; Chu and Turk, 1949). The cation saturation ratio concept proposes ideal proportions of the major exchangeable cations in the soil. However, McLean et al. (1983) have shown that the cation saturation ratio had essentially no impact on yield.

Development of Diagnostic and Prognostic Criteria

Plant analysis may be used for either diagnosis or prognosis. For each purpose results of plant analysis need to be calibrated. Although some promising results were obtained in developing prognostic criteria (Spencer, Jones, and Freney, 1977; Møller Nielsen, 1979a, b), the practical use in routine plant testing programs is limited. Therefore, we focus on the development of diagnostic criteria.

In most plant testing programs the elemental content of plant parts, e.g., fully expanded leaves, is determined to establish the nutritional status of the crop. For evaluation of that nutritional status, diagnostic criteria have been established.

Critical value. When the nutrient concentration in a plant (part) increases, the plant growth rate also increases until the so-called critical level is reached (Figure 3). Concentrations beyond the critical level do not lead to an increase in the growth rate (luxury range), and extremely high indices may even impair growth (toxic range). From an agricultural point of view, the critical level may be a valuable standard for diagnosis of the nutritional plant status (Ulrich and Hills, 1967). For many crops, critical levels have been proposed as a standard for diagnosing the nutritional plant status.

However, the critical level is not a constant; it may differ among crop varieties and is affected by e.g., nutrient interactions, water supply, temperature, dry matter yield level and physiological maturity of the leaf or plant part sampled (Bates, 1971). Moreover, determination of critical levels may lead to inaccurate values. Attempts have been made to overcome these problems by: (1) dividing plant nutrient concentrations into plant status categories (Jones, 1967); (2) defining critical nutrient ranges with the upper limit set at the critical level (Dow and Roberts, 1982); (3) using sufficiency ranges with the lower limit of the sufficient range set at about the critical level (Jones, 1967); and (4) establishing critical levels for different growth stages (Figure 4; see also Tyler and Lorenz, 1962; Pritchard, Doerge, and Thompson, 1995). Despite all these attempts, the critical level has still considerable limitations for its wide use as a diagnostic tool.

An exciting extension has been given by Webb (1972). When sufficient nutrient concentrations of a crop are plotted against crop yield, a skewed spread of points will result. The border of this spread is the maximum crop response to this concentration and is often referred to as the boundary line. For points lower than the boundary line, crop yield is considered to be restricted. The boundary line approach seems a valid way to determine the relationship between the critical level of a nutrient and crop yield. The disadvantages of this approach are: (1) the

FIGURE 4. Interpretation of leaf lettuce midrib tissue NO_3-N concentrations throughout the growth period (Pritchard, Doerge, and Thompson, 1995, with kind permission from Marcel Dekker, Inc., New York, USA).

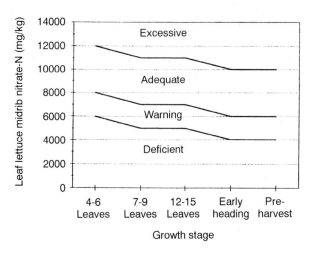

large number of observations required, and (2) the construction of the accurate boundary line.

Nutrient ratio. Nutrient uptake and dry matter accumulation rarely proceed at equal rates in crops. Therefore, concentrations of nutrients expressed on a dry matter basis are generally not constant over time (Lorenz, Tyler, and Fullmer, 1964; Walworth and Sumner, 1988). The concentration of nutrients such as N, P, K, and S in whole plants or plant tissues tends to decrease as dry matter accumulates, while the concentration of Ca and Mg tends to increase or to remain constant (Smith, 1962; Rominger, Smith, and Peterson, 1975; Jiménez et al., 1996). Beaufils (1973) proposed using the nutrient-to-dry matter ratio to eliminate effects of dry matter accumulation.

The nutrient ratio approach has been applied to routine foliar diagnosis and forms part of the Diagnosis and Recommendation Integrated System (DRIS) (Beaufils, 1973). In DRIS, optimal nutrient ratios and the acceptable deviation from these ratios are obtained by collecting nutrient indices from healthy, high-yielding crops. Subsequently, DRIS indices are calculated for each nutrient, giving information about which nutrient is most yield-limiting and also about the order of nutrient requirement (Walworth and Sumner, 1988; Munson and Nelson,

1990). The DRIS was introduced as a universal approach for determining nutrient requirements. Nowadays, its value is debated since it appears a relatively site-specific approach (Beverly, 1993; Baldock and Schulte, 1996). Baldock and Schulte (1996) proposed PASS (Plant Analysis with Standarized Scores) for interpretation of plant analysis. PASS is a combination of the sufficiency range approach and DRIS.

Alternatives. Prevot and Ollagnier (1961a, b) estimated the relative proportions of (interacting) nutrients in plants, which should result in balanced plant growth. Kenworthy (1967, 1973) proposed generating nutrient optima by averaging tissue values of healthy crops rather than by determination of critical values from crop response studies. Møller Nielsen (1971) proposed a system which addresses problems associated with plant physiological age and nutrient interactions. Although the approach is innovative, the amount of data necessary to work out this concept is extremely large and therefore not very promising for wide application in agriculture.

CRITICAL EVALUATION OF SOIL AND PLANT TESTING PROGRAMS

Collection of Soil Samples

Agricultural fields are variable in the horizontal and vertical direction because of natural variation, e.g., soil forming processes (Finke, Bouma, and Stein, 1992), and human influences, e.g., row application of fertilizers (Hofman et al., 1993). Soil sampling schemes should take into account this variability in order to obtain representative analytical data and to develop adequate soil testing programs (Peck and Soltanpour, 1990). Different soil sampling strategies have been proposed to obtain samples which accurately reflect the whole field's nutrient status or parts of it (Kitchen, Havlin, and Westfall, 1990; Mahler, 1990; Entz and Chang, 1991; Blair and Lefroy, 1993; James and Hurst, 1995).

Most current soil testing programs, fertilizer recommendations and fertilizer application techniques aim at one homogeneous application per field. This approach seems inadequate for non-homogeneous fields because it may lead to underdosage or overdosage of fertilizers, resulting in reductions of crop yield and crop quality or in nutrient losses to the environment. When variability is large, knowledge of the spatial variability in combination with site-specific fertilizer application techniques are promising tools to adjust nutrient availability to plant demand and to reduce the risk of losses to the environment (Robert, Rust, and Larson, 1996).

Traditionally, soil samples are taken from the 5 to 30 cm top layer of agricultural soils, mainly because the major portion of the root system is in this layer (De Willigen and Van Noordwijk, 1987). However, crops can take up considerable amounts of nutrients from the subsoil (Kuhlmann and Baumgartel, 1991). This holds true especially for nutrients like K, NO_3 and SO_4 under conditions where the precipitation surplus is small and drainage rarely occurs. Soil testing programs can be improved by estimating the soil's nutrient reserves to a depth related to the rooting zone (Neeteson, 1989).

Most present day soil testing programs aim at collecting one soil sample per year for 'mobile' nutrients and one soil sample per crop rotation (3-6 years) for 'immobile' nutrients. This seems tricky because soil fertility status may show considerable seasonal variation (Espinoza et al., 1991; Carr and Ritchie, 1993). We think that the sampling frequency of present day soil programs is far from sufficient for strategies that aim at fine-tuning of soil nutrient availability to plant demand. Regular soil analysis during the growing season should become an essential part of these strategies, especially for nutrients which are not well buffered in soils.

Collection of Plant Samples

The nutrient content of a plant is not a fixed entity, but varies from month to month, day to day and even from hour to hour as well as between plant organs. Plant sampling schemes should be adapted to this variability in order to be a true and accurate tool for monitoring the crop nutrient concentration (Bolland, 1995). In general, organs that are physiologically young and are subject to rapid changes in nutritional concentration, and organs that have passed full maturity should not be sampled (Bouma, 1983; Jones and Case, 1990; Ernst, 1995). After a period of stress due to possible nutritional deficiency or imbalance, crops develop unusual nutrient concentrations which can lead to serious misinterpretation of the nutritional status. The necessity for standardization of sampling techniques and protocols cannot be overemphasized (Farina, 1994), since current criteria for the interpretation of plant analysis data have been established for well-defined conditions only.

Extraction Procedures

The chemical extraction of a soil or plant sample and subsequent determination of the nutrient concentration in the extract is the basis of soil and plant testing programs. If analytical procedures are carried out under well-defined conditions, reproducibility and accuracy is generally very high. The protocols should at least contain information on sample preparation (drying temperature, duration of drying, mixing), extraction conditions (temperature, soil-to-solution ratio, shaking time, method of separation of soil or plant material from extractant, etc.), and the use of reference or certified soil samples.

Interpretation of Plant Nutrient Concentrations

It is routine in most laboratories to express nutrient concentrations in plant material on a dry matter basis. This has some advantages because dry matter is a measure of crop biomass that will not change much with post harvest treatments. However, it is found that nutrient concentrations expressed on a dry matter basis vary with time; concentrations are mostly high in young plants and decline during ageing (Smith, 1962). It is, therefore, not very useful to define a 'critical' concentration that is required for maximum growth without making a clear reference to the developmental stage at sampling. When nutrient concentrations are expressed on a dry matter basis, the fact that the physiological activity of a nutrient is related to its concentration in the aqueous phase is ignored. More accurate insight into the plant nutritional status may therefore be obtained by expressing concentrations relative to water content. Leigh and Johnston (1983a, b) have shown that leaves of barley adequately supplied with K had more or less constant K concentrations relative to water content throughout vegetative and early reproductive growth, while the concentration based on dry matter declined during growth (Figure 5). This concept is also useful for nitrogen (Leigh and Johnston, 1985).

The total nutrient concentration of a plant part does not always give an indication of the physiological nutrient activity. Analysis of plant sap, as is the case with organ testing, may improve the diagnostic

FIGURE 5. Time-dependent changes in the concentration of K in shoots of barley, expressed on the basis of (A) dry matter and (B) tissue water. Exchangeable K concentrations of the soils were 382 (●) or 55 (▲) mg kg^{-1} dry soil, respectively (Leigh and Storey, 1991, with kind permission from Cambridge University Press, Cambridge, UK).

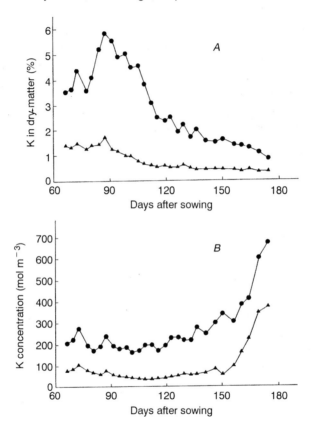

value of the nutrient concentrations obtained. Nutrients in plant organs can be heterogeneously distributed between cell types. This distribution may have implications for the interpretation of nutrient concentrations in plant organs. The nutrient concentration of whole organs has little validity concerning physiological status of the composing cell types. On the intracellular level, nutrients are distributed between the cytoplasm and the vacuole. The cytoplasm and its organelles represent the site of most metabolic processes, and the nutrient composition and the nutrient concentrations should remain more or less constant in a certain developmental stage. Physiological nutrient shortages of N, P and K appear to be largely influenced by cytoplasmic nutrient concentrations (Mengel and Kirkby, 1987; Marschner, 1995). Nutrients present in the vacuole, such as K, contribute to the osmotic potential (Leigh and Wyn Jones, 1984) but have no unique or essential role in the vacuole. Providing that other solutes are available to maintain turgescence (Pitman, Mowat, and Nair, 1971), the concentration of any nutrient in the vacuole, and thus in the cells or plant organs as a whole, may vary with supply. This explains why the critical concentration of K in plant tissue is affected by the availability of cations like Na, Mg and Ca (Leigh and Storey, 1991).

From the above considerations it is clear that total nutrient concentrations in plant organs have little, if any, plant physiological meaning. Despite this, total nutrient concentrations in plant testing programs that are properly calibrated may be correlated to the plant nutritional status.

Interpretation of Soil Extractable Nutrients

In the soil, nutrients are retained in many chemical forms. The absolute or relative amount of a nutrient that is released during extraction depends on the total amount present, on the distribution over the different chemical binding forms, and on the extracting power of the extractant. There is generally an enormous discrepancy between the amount of nutrient extracted and actual nutrient uptake by crops. Furthermore, nutrient uptake by the same crop may vary between years and different crops may take up different amounts of nutrients from the same soil (Yerokun and Christenson, 1990; Schoenau and Huang, 1991; Smith and Li, 1993). The mode of action of many soil extractants when brought into contact with the soil sample is still largely unknown. At present, many extractants are in use for the assessment of a single-element fertility status of the soils. When different extractants are applied to the same soil, the amount of extractable nutrients may differ enormously (Matejovic and Durackova, 1994).

Crop Yield

Soil and plant testing programs are correlated, calibrated and interpreted with crop yield as the determinant (Dahnke and Olson, 1990;

Munson and Nelson, 1990; Blair and Lefroy, 1993). When crop yields of many field trials are plotted against a single independent variable, e.g., soil fertility or plant nutrient content, a spread of points will result because under field conditions other independent variables may also change (Webb, 1972; Walworth, Letzsch, and Sumner, 1986). The maximum response to an independent variable is the border of the spread of points, but the majority of the fields have crop yields below this border. In all these cases, factors other than nutrient availability have determined actual crop yield. The dependence of crop yield on other factors indicates that soil and plant testing programs need more background research on the contribution of these factors to crop yield before fertilization can be optimized on a field scale.

Fertilizer Application and Crop Response

There is a relationship between soil fertility status, fertilizer application rate and crop response. To achieve the desired crop yield, the fertilizer application rate on soils with a low fertility status should be higher than on soils with a higher fertility status. There is, however, no single relationship which can be used to describe changes in plant nutrient concentration upon addition of fertilizer. Thus, foliar diagnosis alone cannot be used to determine how much fertilizer to add, or to predict accurately crop response to added fertilizer in any given situation. These relationships are largely affected by non-nutritional factors, and further in-depth research is necessary before they can be incorporated successfully in future fertilizer recommendations (Walworth and Sumner, 1988; Beverly, 1993; Marschner, 1995).

Environmental Side-Effects Related to Soil Testing Programs

Most soil testing programs have been calibrated under a wide diversity of field conditions. These programs are currently a useful practical tool to match the fertilizer application rate with the soil fertility status, the crop yield target, and crop quality, and at the same time to achieve the sufficient soil fertility status. The recommended application rates will generally lead to relatively high nutrient-use efficiencies. When fertilizer application exceeds the recommended rates, nutrients will certainly be left in the soil profile at harvest. The 'mobile' nutrients are then subject to leaching during a period of precipitation surplus, while the 'immobile' nutrients may remain in the rooting zone of the soil profile and increase the soil fertility status. When subsequent fertilizer application rates are not adjusted to the enhanced soil fertility status, the risk of nutrient losses through leaching, denitrification or surface run-off may increase accordingly.

Most fertilizer recommendation schemes have been developed from fertilizer field trials on well-defined fields where water-soluble single nutrient fertilizers with a well-known composition had been broadcast and further incorporated in the soil just before sowing or planting. If these experimental boundary conditions are not taken into account during the practical application of fertilizers, fertilization according to the recommendation schemes may lead to considerable environmental side-effects or decreased nutrient availability. Use of multi-nutrient fertilizer generally leads to under or overdosage of more than one nutrient.

To increase nutrient use efficiency and to alleviate side-effects of fertilization, soil testing programs should also provide information on the impact of choice of fertilizer (e.g., inorganic versus organic forms) and on timing and methods of application. Application of fertilizers long before planting or sowing may result in leaching and/or denitrification losses (Addiscott, Whitmore, and Powlson, 1991). Surface application of ammonium-containing fertilizers on carbonate-containing soils and of urea-containing fertilizers on all soil types may lead to ammonia volatilization. Injection or direct incorporation of these fertilizers in the top layer of soil decreases ammonia volatilization (Hargrove, 1988). Broadcast application of water-soluble P and K fertilizers on P- or K-fixing soils, respectively, may result in decreased P or K availability to crops (Tingre et al., 1992). Banding just before sowing or planting may then improve the plant availability of applied P and K (Knittel, 1988).

EVALUATION

Thus far, the basis of most soil and plant testing programs is the statistical relationship between the concentration of one or more extracted plant nutrients and crop response. Since many factors determine crop response, the relationships are frequently not very strong and, moreover, site and crop specific. Despite the empirical and site-specific approach of most plant and soil testing programs, they are still the best tools available to optimize fertilizer strategies under these specific conditions. The testing programs are effective, especially when (soil) nutrient availability is the major factor restricting crop response.

In European and North-American regions, agriculture is confronted with an increasing number of agricultural, environmental, economic and legislative boundary conditions that restrict fertilizer use. To satisfy these demands and constraints, fertilizer strategies like Integrated Nutrient Management (Van Erp and Oenema, 1993) and Balanced Fertilization (Steen, 1996) have been proposed. Integrated Nutrient Management (INM) seems most realistic from an agricultural point of view and aims at monitoring and steering of nutrient flows in the soil-plant system. Computorized crop growing models and quantitative risk analysis techniques may be helpful tools to estimate fertilizer requirements and the probability of responses. The use of multi-nutrient extractants should be promoted to reduce the number of soil analyses.

For INM, a more scientific approach of soil and plant testing programs will be unavoidable. Therefore, future soil and plant testing programs should focus on the extraction and determination of nutrient (fractions) that are related to relevant soil and plant processes and can be used in crop growing models.

REFERENCES

Addiscott, T.M., A.P. Whitmore, and D.S. Powlson. (1991). *Farming, Fertilizers and the Nitrate Problem.* Wallingford, UK: CAB International.

Anonymous. (1987). Decree on the Use of Animal Manures (In Dutch). In *Staatsblad van het Koninkrijk der Nederlanden No. 386.* The Hague, The Netherlands: SDU Publishers, 8 p.

Baldock, J.O. and E.E. Schulte. (1996). Plant analysis with standardized scores combines DRIS and sufficiency range approaches for corn. *Agronomy Journal* 88: 448-456.

Barlett, R. and B. James. (1980). Studying dried, stored soil samples–some pitfalls. *Soil Science Society of America Journal* 44: 721-724.

Bar-Akiva, A. (1971). Functional aspects of mineral nutrients in use for the evaluation of plant nutrient. In *Recent Advances in Plant Nutrition*, ed. R.M. Samish. New York, USA: Gordon and Breach Science Publishers, pp. 115-142.

Bar-Akiva, A. (1984). Substitutes for benzidine as N-donors in the peroxidase assay for rapid diagnosis of iron in plants. *Communications in Soil Science and Plant Analysis* 15: 929-934.

Barber, S.A. (1984). *Soil Nutrient Bioavailability.* New York, USA: John Wiley & Sons.

Bates, T.E. (1971). Factors affecting critical nutrient concentrations in plants and their evaluations: a review. *Soil Science* 112: 116-130.

Bear, F.E. and S.J. Toth. (1948). Influence of Ca on availability of other soil cations. *Soil Science* 65: 69-74.

Beaufils, E.R. (1973). *Diagnosis and Recommendation Integrated System (DRIS). A General Scheme for Experimentation and Calibration Based on Principles Developed from Research in Plant Nutrition.* Pietermaritzburg, South Africa: University of Natal, Soil Science Bull. No. 1.

Beverly, R.B. (1993). Re-evaluation reveals weaknesses of DRIS and sufficiency range diagnoses for wheat, corn and alfalfa. *Communications in Soil Science and Plant Analysis* 24: 487-501.

Blair, G. and R. Lefroy. (1993). Interpretation of soil tests: a review. *Australian Journal of Experimental Agriculture* 33: 1045-1052.

Bolland, M.D.A. (1995). Variation of plant-test phosphorus for individual lupin and wheat tops. *Communications in Soil Science and Plant Analysis* 26: 2511-2517.

Bouma, D. (1983). Diagnosis of mineral deficiencies using plant tests. In *Encyclopedia of Plant Physiology, Vol. 15A: Inorganic Plant Nutrition*, eds. A. Läuchli and R.L. Bieleski. Berlin, Germany: Springer-Verlag, pp. 120-146.

Breimer, T. (1982). Environmental factors and cultural measures affecting the nitrate content of spinach. *Fertilizer Research* 3: 191-292.

Brown, P.H., R. Welch, E.E. Cary, and R.T. Checkai. (1987). Beneficial effects of nickel on plant growth. *Journal of Plant Nutrition* 10: 2125-2135.

Bussink, D.W. (1992). Ammonia volatilization from grassland receiving nitrogen fertilizer and rotationally grazed by dairy cattle. *Fertilizer Research* 33: 257-265.

Bussink, D.W. (1994). Relationships between ammonia volatilization and nitrogen fertilizer application rate, intake and excretion of herbage nitrogen by cattle on grazed swards. *Fertilizer Research* 38: 111-121.

Carr, S.J. and G.S.P. Ritchie. (1993). Temporal variations in potassium chloride extractable aluminium, sodium and soil pH, and the effects on interpretation of subsoil aluminium toxicity in yellow earths. *Communications in Soil Science and Plant Analysis* 24: 513-523.

Cate, R.B. and L.A. Nelson. (1965). *A Rapid Method for Correlation of Soil Test Analyses with Plant Response Data.* Raleigh, North Carolina, USA: North Carolina State University, Int. Soil Testing Series Tech. Bull. No. 1.

Cate, R.B. and L.A. Nelson. (1971). A simple statistical procedure for partitioning soil test correlation data into two classes. *Soil Science Society of America Proceedings* 35: 658-660.

Chu, T.S. and L.M. Turk. (1949). *Growth and Nutrition of Plants as Affected by Degree of Base Saturation of Different Types of Clay Minerals.* Michigan Agricultural Experimental Station Tech. Bull. 214.

Corey, R.B. (1987). Soil test procedures: correlation. In *Soil Testing: Sampling, Correlation, Calibration, and Interpretation*, Special Publication 21, ed. J.R. Brown. Madison, Wisconsin, USA: Soil Science Society of America, pp. 15-22.

Dahnke, W.C. and R.A. Olson. (1990). Soil test correlation, calibration and recommendations. In *Soil Testing and Plant Analysis, 3rd edition*, ed. R.L. Westermann. Madison, Wisconsin, USA: Soil Science Society of America, pp. 45-71.

Dahlgren, R.A. (1993). Comparison of soil solution extraction procedures: effect on solute chemistry. *Communications in Soil Science and Plant Analysis* 24: 1783-1794.

De Willigen, P. and M. Van Noordwijk. (1987). *Roots, Plant Production and Nutrient Use Efficiency*. Doctoral Thesis, Wageningen Agricultural University, Wageningen, The Netherlands.

Dow, A.I. and S. Roberts. (1982). Proposal: critical nutrient ranges for crop diagnosis. *Agronomy Journal* 74: 401-403.

EC. (1991). *Council Directive Concerning the Protection of Waters Against Pollution Caused by Nitrates from Agricultural Sources (91/676/EEG)*. Official Journal of the European Communities, nr. L 375/1. Luxemburg: Bureau voor officiele publikaties der Europese Gemeenschappen.

Entz, T. and C. Chang. (1991). Evaluation of soil sampling schemes for geostatistical analyses: a case study for soil bulk density. *Canadian Journal of Soil Science* 71: 165-176.

Epstein, E. (1965). Mineral metabolism. In *Plant Biochemistry*, eds. J. Bonner and J.E. Varner. London, UK: Academic Press, pp. 438-466.

Epstein, E. (1972). *Mineral Nutrition of Plants: Principles and Perspectives*. New York, USA: John Wiley & Sons.

Ernst, W.H.O. (1995). Sampling of plant material for chemical analysis. *The Science of the Total Environment* 176: 15-24.

Espinoza, J.E., L.R. McDowell, N.S. Wilkinson, J.H. Conrad, and F.G. Martin. (1991). Monthly variations of forage and soil minerals in Central Florida: I. Macro-minerals. *Communications in Soil Science and Plant Analysis* 22: 1123-1136.

Evangelou, V.P., J. Wang, and R.E. Phillips. (1994). New developments and perspectives on soil potassium quantity/intensity relationships. *Advances in Agronomy* 52: 173-227.

FAO. (1995). *Fertilizers 1994*. Rome, Italy: FAO Statistics Series No. 126.

Farina, M.P.W. (1994). Improving the quality of soil and plant samples. *Communications in Soil Science and Plant Analysis* 25: 781-797.

Finke, P.A., J. Bouma, and A. Stein. (1992). Measuring field variability of disturbed soil for simulation. *Soil Science Society of America Journal* 56: 187-192.

Fitts, J.W. (1955). Using soil tests to predict a probable response from fertilizer application. *Better Crops with Plant Food* 39: 17-20.

Fixen, P.E. and J.H. Grove. (1990). Testing soils for phosphorus. In *Soil Testing and Plant Analysis, 3rd edition*, ed. R.L. Westermann. Madison, Wisconsin, USA: Soil Science Society of America, pp. 141-180.

Fluck, R.C., ed. (1992). *Energy in Farm Production*. Amsterdam, The Netherlands: Elsevier Science Publishers.

Granli, T. and O.C. Bøckman. (1994). Nitrous oxide from agriculture. *Norwegian Journal of Agricultural Sciences*, Supplement 12.

Haby, V.A., M.P. Russelle, and E.O. Skogley. (1990). Testing soils for potassium, calcium and magnesium. In *Soil Testing and Plant Analysis, 3rd edition*, ed. R.L. Westermann. Madison, Wisconsin, USA: Soil Science Society of America, pp. 181-227.

Hargrove, W.L. (1988). Soil, environmental, and management factors influencing ammonia volatilization under field conditions. In *Ammonia Volatilization from Urea Fertilizers*, eds. B.R. Bock and D.E. Kissel. Muscle Shoals, Alabama, USA: National Fertilizer Development Center, Tennessee Valley Authority, pp. 17-36.

Hauser, G.F. (1973). *Guide to the Calibration of Soil Tests for Fertilizer Recommendations*. Rome, Italy: FAO Soils Bull. 18.

Hernandez, L.E., A.M. Ramon, R.O. Carpena-Ruiz, and A. Garate. (1995). Evaluation of nitrate nutrition indexes in maize leaves: metabolic nitrate, total nitrate content and nitrate reductase activity. *Journal of Plant Nutrition* 18: 869-887.

Hofman, G., P. Verstegen, P. Demyttenaere, M. Van Meirvenne, P. Delanote, and G. Ampe. (1993). Comparison of row and broadcast N application on N efficiency and yield of potatoes. In *Optimization of Plant Nutrition*, eds. M.A.C. Fragoso and M.L. Van Beusichem. Dordrecht, The Netherlands: Kluwer Academic Publishers, pp. 359-365.

Holford, I.C.R. (1991). Comments on intensity-quantity aspects of soil phosphorus. *Australian Journal of Soil Research* 29: 11-14.

Holford, I.C.R. and A.D. Doyle. (1992). Influence of intensity/quantity characteristics of soil phosphorus test on their relationships to phosphorus responsiveness of wheat under field conditions. *Australian Journal of Soil Research* 30: 343-356.

Houba, V.J.G., I. Novozamsky, and J.J. Van der Lee. (1989). Some aspects of determinations of nitrogen fractions in soil extracts. *VDLUFA-Schriftenreihe* 30: 305-312.

Houba, V.J.G., I. Novozamsky, T.M. Lexmond, and J.J. Van der Lee. (1990). Applicability of 0.01 M $CaCl_2$ as a single extraction solution for the assessment of the nutrient status of soils and other diagnostic purposes. *Communications in Soil Science and Plant Analysis* 21: 2281-2290.

Houba, V.J.G., I. Novozamsky, and J.J. Van der Lee. (1995). Influence of storage of plant samples on their chemical composition. *The Science of the Total Environment* 176: 73-79.

IFA, IFDC, and FAO. (1994). *Fertiliser Use by Crops*. Rome, Italy: ESS/MISC/1994/4.

IFA. (1995). *Fertilizer Indicators; Graphs and Diagrams*. Paris, France: International Fertilizer Industry Association.

Isermann, K. (1990). Share of agriculture in nitrogen and phosphorus emissions into the surface waters of Western Europe against the background of their eutrophication. *Fertilizer Research* 26: 253-269.

James, D.W. and K.L. Wells. (1990). Soil sample collection and handling: technique based on source and degree of field variability. In *Soil Testing and Plant Analysis, 3rd edition*, ed. R.L. Westermann. Madison, Wisconsin, USA: Soil Science Society of America, pp. 25-44.

James D.W. and R.L. Hurst. (1995). Soil sampling techniques for band-fertilized, no-till fields with Monte Carlo simulations. *Soil Science Society of America Journal* 59: 1768-1772.

Jiménez, M.P., D. Effrón, A.M. de la Horra, and R. Defrieri. (1996). Foliar potassium, calcium, magnesium, zinc and manganese content in soybean cultivars at different stages of development. *Journal of Plant Nutrition* 19: 807-816.

Jones, J.B. (1967). Interpretation of plant analysis for several agronomic crops. In

Soil Testing and Plant Analysis, II: Plant Analysis, Special Publication 2, ed. M. Stelly. Madison, Wisconsin, USA: Soil Science Society of America, pp. 49-58.

Jones, J.B. and V.W. Case. (1990). Sampling, handling, and analysing plant tissue samples. In *Soil Testing and Plant Analysis, 3rd edition*, ed. R.L. Westermann. Madison, Wisconsin, USA: Soil Science Society of America, pp. 389-428.

Jones, D.L. and A.C. Edwards. (1993). Evaluation of polysulfone hollow fibres and ceramic suction samplers as devices for the in situ extraction of soil solution. *Plant and Soil* 150: 157-165.

Kenworthy, A.L. (1967). Plant analysis and interpretation of analysis for horticulture crops. In *Soil Testing and Plant Analysis, II: Plant Analysis*, Special Publication 2, ed. M. Stelly. Madison, Wisconsin, USA: Soil Science Society of America, pp. 59-75.

Kenworthy, A.L. (1973). Leaf analysis as an aid in fertilizing orchards. In *Soil Testing and Plant Testing*, eds. L.M. Walsh and J.D. Beaton. Madison, Wisconsin, USA: Soil Science Society of America, pp. 381-392.

Kitchen, N.R., J.L. Havlin, and D.G. Westfall. (1990). Soil sampling under no-till banded phosphorus. *Soil Science Society of America Journal* 54: 1661-1665.

Knittel, H. (1988). Placement of soil fertilisers in agricultural crops: a review. *Proceedings of The Fertiliser Society* 273: 3-20.

Koops, J.G., O. Oenema, and M.L. Van Beusichem. (1996). Denitrification in the top and subsoil of grassland on peat soil. *Plant and Soil* 184: 1-10.

Kuhlmann, H. and G. Baumgartel. (1991). Potential importance of the subsoil for the P and Mg nutrition of wheat. *Plant and Soil* 137: 259-266.

Lawrence, G.B. and M.B. David. (1996). Chemical evaluation of soil solution in acid forests soils. *Soil Science* 161: 298-313.

Leigh, R.A. and A.E. Johnston. (1983a). Concentration of potassium in the dry matter and tissue water of field-grown spring barley and their relationships to grain yield. *Journal of Agricultural Science (Cambridge)* 101: 675-685.

Leigh, R.A. and A.E. Johnston. (1983b). The effects of fertilizers and drought on the concentrations of potassium in the dry matter and tissue water of field-grown barley. *Journal of Agricultural Science (Cambridge)* 101: 741-748.

Leigh, R.A. and A.E. Johnston. (1985). Nitrogen concentration in field-grown spring barley: an examination of the usefulness of expressing concentrations on the basis of tissue water. *Journal of Agricultural Science (Cambridge)* 105: 397-406.

Leigh, R.A. and R.G. Wyn Jones. (1984). A hypothesis relating the critical potassium concentration for growth to the distribution and functions of this ion in the plant cell. *The New Phytologist* 97: 1-13.

Leigh, R.A. and R. Storey. (1991). Nutrient compartmentation in cells. In *Plant Growth, Interactions with Nutrition and Environment*, eds. J.R. Porter and D.W. Lawlor. Cambridge, UK: Cambridge University Press, pp. 33-54.

Lorenz, O.A., K.B. Tyler, and F.S. Fullmer. (1964). Plant analysis for determining the nutritional status of potatoes. In *Plant Analysis and Fertilizer Problems IV*, eds. C. Bould, P. Prevot, and J.R. Magness. New York, USA: W.F. Humphrey Press, pp. 226-240.

Lorenz, S.E., R.E. Hamon, and S.P. McGrath. (1994). Differences between soil solutions obtained from rhizosphere and non-rhizosphere soils by water displacement and soil centrifugation. *European Journal of Soil Science* 45: 431-438.

Mahler, R.L. (1990). Soil sampling fields that have received banded fertilizer applications. *Communications in Soil Science and Plant Analysis* 21: 13-16.

Marschner, H. (1995). *Mineral Nutrition of Higher Plants, 2nd edition*. London, UK: Academic Press.

Martin-Prével, P., J. Gagnard, and P. Gautier, eds. (1987). *Plant Analysis as a Guide to the Nutrient Requirements of Temperate and Tropical Crops*. New York, USA: Lavoisier Publishing.

Matejovic, I. and A. Durackova. (1994). Comparison of Mehlich 1-, 2-and 3-, Calciumchloride-, Bray-, Olsen-, Enger- and Schachtschabel-extractants for determinations of nutrients in two soil types. *Communications in Soil Science and Plant Analysis* 25: 1289-1302.

McLachlan, K.D. (1982). Leaf acid phosphatase activity and the phosphorus status of field-grown wheat. *Australian Journal of Agricultural Research* 33: 453-464.

McLean, E.O., R.C. Hartwig, D.J. Eckert, and G.B. Triplett. (1983). Basic cation saturation ratios as a basis for fertilizing and liming agronomic crops. II. Field studies. *Agronomy Journal* 75: 635-639.

Menon, R.G., S.H. Chien, and W.J. Chardon. (1997). Iron oxide-impregnated filter paper (Pi test): II. A review of its application. *Nutrient Cycling in Agroecosystems* 47: 7-18.

Mengel, K. and E.A. Kirkby. (1987). *Principles of Plant Nutrition, 4th edition*. Berne, Switzerland: International Potash Institute.

Meyer, W.L. and P.A. Arp. (1994). Exchangeable cations and cation exchange capacity of forest soil samples: effects of drying, storage and horizon. *Canadian Journal of Soil Science* 74: 421-429.

Mitscherlich, E.A. (1928). The second approximation of the effect of growth factors. *Zeitschrift für Pflanzenernährung* 12: 273-282.

Møller Nielsen, J. (1971). Diagnosis and control of nutritional disorders in cereals based on inorganic tissue analysis. In *Recent Advances in Plant Nutrition*, ed. R.M. Samish. New York, USA: Gordon and Breach Science Publishers, pp. 63-73.

Møller Nielsen, J. (1979a). Evaluation and control of the nutritional status of cereals. IV. Quantity of final yield controlled by nutrient therapy. *Plant and Soil* 52: 229-244.

Møller Nielsen, J. (1979b). Evaluation and control of the nutritional status of cereals. IV. Quality of final yield controlled by nutrient therapy. *Plant and Soil* 52: 245-268.

Munson, R.D. and W.L. Nelson. (1990). Principles and practices of plant analysis. In *Soil Testing and Plant Analysis, 3rd edition*, ed. R.L. Westermann. Madison, Wisconsin, USA: Soil Science Society of America, pp. 359-387.

Neeteson, J.J. (1989). Evaluation of the performance of three advisory methods for nitrogen fertilization of sugar beet and potatoes. *Netherlands Journal of Agricultural Science* 37: 143-155.

Nelson, L.A. and R.L. Anderson. (1977). Partitioning of soil test–crop response probability. In *Soil Testing: Correlating and Interpreting the Analytical Results*, Special Publication 29, ed. M. Stelly. Madison, Wisconsin, USA: American Society of Agronomy, Crop Science Society of America, Soil Science Society of America, pp. 19-38.

Nye, P.H. and P.B. Tinker. (1977). *Solute Movement in the Soil-Root System*. Los Angeles, California, USA: University of California Press.

Oenema, O. (1990). Calculated rates of soil acidification of intensively used grassland in the Netherlands. *Fertilizer Research* 26: 217-228.

Peck, T.R. and P.N. Soltanpour. (1990). The principles of soil testing. In *Soil Testing and Plant Analysis, 3rd edition*, ed. R.L. Westermann. Madison, Wisconsin, USA: Soil Science Society of America, pp. 1-9.

Pitman, M.G., J. Mowat, and H. Nair. (1971). Interpretation of the processes for accumulation of salt and sugar in barley roots. *Australian Journal of Biological Sciences* 24: 619-631.

Prevot, P. and M. Ollagnier. (1961a). Law of the Minimum and balanced mineral nutrition. In *Plant Analysis and Fertilizer Problems*, Publication 8, ed. W. Reuther. Washington, D.C., USA: American Institute of Biological Sciences, pp. 257-277.

Prevot, P. and M. Ollagnier. (1961b). Foliar diagnosis: reciprocal relationships of certain mineral elements (In French). In *Advances in Horticultural Science*, ed. J.C. Garnaud. Oxford, UK: Pergamon Press, pp. 217-228.

Pritchard, K.H., T.A. Doerge, and T.L. Thompson. (1995). Evaluation of in season nitrogen tissue tests for drip irrigated leaf and romaine lettuce. *Communications in Soil Science and Plant Analysis* 26: 237-257.

Qian, P., J.J. Schoenau, and W.Z. Huang. (1992). Use of ion exchange membranes in routine soil testing. *Communications in Soil Science and Plant Analysis* 23: 1791-1804.

Raven, K.P. and Hossner L.R. (1994). Sorption and desorption quantity-intensity parameters related to plant-available phosphorus. *Soil Science Society of America Journal* 58: 405-410.

Rechcigl, J.E., G.G. Payne, and C.A. Sanchez. (1992). Comparison of various soil drying techniques on extractable nutrients. *Communications in Soil Science and Plant Analysis* 23: 2347-2363.

Rezaian, S., E.A. Hanlon, C.A. Sanchez, and J.A. Cornell. (1992). Optimization of solution-soil ratio and shaking times of the Mehlich-III soil tests on Histosols. *Communications in Soil Science and Plant Analysis* 23: 17-20.

Robert, P.C., R.H. Rust, and W.E. Larson, eds. (1996). *Proceedings of The Third International Conference on Precision Agriculture*. Madison, Wisconsin, USA: American Society of Agronomy, Crop Science Society of America, Soil Science Society of America.

Rominger, R.S., D. Smith, and L.A. Peterson. (1975). Changes in elemental concentrations in alfalfa herbage at two soil fertility levels with advance in maturity. *Communications in Soil Science and Plant Analysis* 6: 163-180.

Rubaek, G.H. and E. Sibbesen. (1993). Resin extraction of labile soil organic phosphorus. *Journal of Soil Science* 44: 467-468.

Schoenau, J.J. and W.Z. Huang. (1991). Anion-exchange membrane, water and sodium bicarbonate extractions as soil tests for phosphorus. *Communications in Soil Science and Plant Analysis* 22: 465-492.

Schofield, R.K. (1955). Can a precise meaning be given to 'available' soil phosphorus? *Soils and Fertilizers* 28: 373-375.

Slangen, J.H.G., H.H.H. Titulaer, and C.A.E. Rijkers. (1989). Nitrogen fertiliser recommendations with the KNS system for iceberg lettuce (*Lactuca sativa* L., var. capita) in field cropping. *VDLUFA-Schriftenreihe* 28: 251-256.

Smaling, E.M.A. (1993). The soil nutrient balance: an indicator of sustainable agriculture in sub-Saharan Africa. *Proceedings of The Fertiliser Society* 340: 1-18.

Smith, K.A. and S.X. Li. (1993). Estimation of potentially mineralisable nitrogen in soil by KCl extraction. I. Comparison with pot experiments. *Plant and Soil* 157: 164-174.

Smith, P.F. (1962). Mineral analysis of plant tissues. *Annual Review of Plant Physiology* 13: 81-108.

Spencer, K., M.B. Jones, and J.R. Freney. (1977). Diagnostic indices for sulphur status of subterranean clover. *Australian Journal of Agricultural Research* 28: 401-412.

Steen, I. (1996). Putting the concept of environmentally balanced fertilizer recommendations into practice on the farm. *Fertilizer Research* 43: 235-240.

Tingre, P.G., H.N. Rawankar, V.A. Deshmukh, and S.M. Bhoyar. (1992). Effect of lime content on phosphorus fixation in soils. *Journal of Soils and Crops* 2: 106-107.

Tyler, K.B. and O.A. Lorenz. (1962). Diagnosing nutrient needs in vegetables. *Better Crops with Plant Food* 46: 6-13.

Ulrich, A. and F.J. Hills. (1967). Principles and practices of plant analysis. In *Soil Testing and Plant Analysis, II: Plant Analysis*, Special Publication 2, ed. M. Stelly. Madison, Wisconsin, USA: Soil Science Society of America, pp. 11-23.

Van Erp, P.J. and O. Oenema. (1993). Towards an integrated nutrient management. *Proceedings of The Fertiliser Society* 345: 1-32.

Van Reuler, H. (1996). *Nutrient Management over Extended Cropping Periods in the Shifting Cultivation System of South-West Côte d'Ivoire.* Doctoral Thesis, Wageningen Agricultural University, Wageningen, The Netherlands.

Velthof, G.L., A.B. Brader, and O. Oenema. (1996). Seasonal variations in nitrous oxide losses from managed grasslands in the Netherlands. *Plant and Soil* 181: 263-274.

VROM and LNV. (1995). *Integral Note on Policy on Manure and Ammonia* (In Dutch). The Hague, The Netherlands: Ministerie van VROM en Ministerie van LNV, 36 p.

Waggoner, P.E. and W.A. Norvell. (1979). Fitting the Law of the Minimum to fertilizer applications and crop yields. *Agronomy Journal* 71: 352-354.

Walworth, J.L., W.S. Letzsch, and M.E. Sumner. (1986). Use of boundary lines in establishing diagnostic norms. *Soil Science Society of America Proceedings* 50: 123-128.

Walworth, J.L. and M.E. Sumner. (1988). Foliar diagnosis: a review. In *Advances in Plant Nutrition, Vol. 3*, eds. P.B. Tinker and A. Läuchli. New York, USA: Praeger Publishers, pp. 193-241.

Webb, R.A. (1972). Use of the boundary line in the analysis of biological data. *Journal of Horticultural Science* 47: 309-319.

World Bank. (1992). *World Development Report 1992. Development and the Environment.* Washington D.C., USA: World Bank, 308 p.

Yanai, J., D.J. Linehan, D. Robinson, I.M. Young, C.A. Hackett, K. Kyuma, and T. Kosaki. (1996). Effects of inorganic nitrogen application on the dynamics of the soil solution composition in the root zone of maize. *Plant and Soil* 180: 1-9.

Yerokun, O.A. and D.R. Christenson. (1990). Relating high soil test phosphorus concentrations to plant phosphorus uptake. *Soil Science Society of America Journal* 54: 796-799.

SUBMITTED: 03/15/97
ACCEPTED: 07/15/97

Comparative Assessment of the Efficacy of Various Nitrogen Fertilizers

Franz Wiesler

SUMMARY. The efficiency of N fertilizers is usually poor; often less than 50% of the applied N is taken up by the crop. This review focuses on various N fertilizers with respect to the significance of different N loss pathways, namely (i) ammonia volatilization, (ii) dinitrogen and nitrogen oxide emissions, and (iii) nitrate leaching. Further, the significance of biological N immobilization, ammonium fixation and, finally, the impact of nitrate vs. ammonium uptake on crop yield are also discussed. The reviewed literature shows that N fertilizers may differ markedly in their susceptibility to losses. There is, however, considerable scope to improve N efficiency of each N source by proper N management practices. *[Article copies available for a fee from The Haworth Document Delivery Service: 1-800-342-9678. E-mail address: getinfo@ haworthpressinc.com]*

KEYWORDS. Ammonia, ammonium, denitrification, efficiency, fertilizer, immobilization, leaching, nitrate, nitrogen, recovery, volatilization

ABBREVIATIONS. For various N fertilizers: AA, anhydrous ammonia; ABC, ammonium bicarbonate; ACl, ammonium chloride; AN, am-

Franz Wiesler, Institute of Plant Nutrition, University of Hannover, Herrenhäuser Straße 2, D-30419 Hannover, Germany (E-mail: Wiesler@mbox.pflern.uni-hannover.de).

The author would like to thank Prof. Dr. W. J. Horst (Institute of Plant Nutrition, Hannover) and Dr. R. Härdter (Intern. Potash Institute, Kassel) for valuable comments and critical reading of the manuscript. Prof. Dr. H. Kuhlmann (Norsk Hydro Agrar) and Dr. W. Zerulla (BASF Ludwigshafen) generously provided statistical data.

[Haworth co-indexing entry note]: "Comparative Assessment of the Efficacy of Various Nitrogen Fertilizers." Wiesler, Franz. Co-published simultaneously in *Journal of Crop Production* (Food Products Press, an imprint of The Haworth Press, Inc.) Vol. 1, No. 2 (#2), 1998, pp. 81-114; and: *Nutrient Use in Crop Production* (ed: Zdenko Rengel) Food Products Press, an imprint of The Haworth Press, Inc., 1998, pp. 81-114. Single or multiple copies of this article are available for a fee from The Haworth Document Delivery Service [1-800-342-9678, 9:00 a.m. - 5:00 p.m. (EST). E-mail address: getinfo@ haworthpressinc.com].

monium nitrate; AS, ammonium sulfate; CaN, calcium nitrate; DAP, diammonium phosphate; KN, potassium nitrate; MAP, monoammonium phosphate; NaN, sodium nitrate; SCU, sulfur-coated urea; UAN, urea-ammonium nitrate solution; U-Ca, urea-$CaCl_2$ solution; U-P, urea-urea phosphate granules; USG, urea supergranules.

Time of N application: BP, before planting; P, at planting; AP, after planting.

Technique of N application: BC, broadcast; BaP, band placement; PP, point placement; I, incorporation; S, surface application.

Inhibitors: NI, nitrification inhibitors [DCD, dicyandiamide; ECC, encapsulated calcium carbide; NP, nitrapyrin]; UI, urease inhibitors [nBTPT, N-(n-butyl)thiophosphorictriamide; PPD, phenylphosphorodiamidate].

CROP YIELD, NITROGEN UPTAKE AND FERTILIZER N RECOVERY AS INFLUENCED BY THE APPLICATION OF VARIOUS N FERTILIZERS

Nitrogen (N) is the most limiting nutrient for crop production in many of the world's agricultural areas. To meet the food and fiber needs of a growing world population, global use of manufactured N fertilizers increased largely, by a factor of about eight, during the past four decades (Bumb, 1995). However, the efficiency of fertilizer N is frequently low, with often less than 50% of the applied N taken up by the crop (Craswell and Godwin, 1984; Strong, 1995). This may cause severe yield limitations where there is a lack of N fertilizers and may increase the risk of environmental pollution of both air and water, particularly where high N fertilizer doses are applied to achieve maximum yields. The comparative effectiveness of different N fertilizer types is shown in Table 1, summarizing studies in which crop yield, nitrogen uptake and fertilizer N recovery by various crops were determined. In all experiments, dry matter yield responded positively to applied N fertilizer. The N recovery from applied fertilizer was estimated by two approaches, the difference method (apparent recovery, for definition see Table 1) and tracer studies using ^{15}N-labeled fertilizers. Generally, the isotope dilution method gives lower recoveries than does the difference method. For a critical comparison of both methods the reader is referred to Jenkinson, Fox, and Rayner (1985) and Rao et al. (1992).

Table 1 shows that recovery of fertilizer N may vary greatly between experiments, averages over treatment ranges being 25-83% (apparent recovery), 25-64% (^{15}N recovery in plants) and 12-44% (^{15}N recovery in soil). Likewise, the percentage of ^{15}N not recovered in the plant/soil system varied between 14 and 64% of applied N. When comparing different N sources, the majority of experiments with upland crops indicate that surface-applied urea gives lower yields and lower N recoveries than ammonium nitrate. The efficacy of ammonium sulfate lays somewhere between urea and ammonium nitrate. There are, however, also studies showing no differences or even larger yields and improved recoveries of fertilizer N when urea was applied instead of nitrate-containing fertilizers (Chancy and Kamprath, 1982; Christensen and Meints, 1982). In some cases amending urea or ammonium fertilizers with nitrification or urease inhibitors had beneficial effects (Chancy and Kamprath, 1982; Watson et al., 1994), whereas no positive effects could be observed in other experiments (Blackmer and Sanchez, 1988). When urea was banded instead of broadcast, N recovery was improved to a level comparable to ammonium nitrate (Malhi, 1995).

As repeatedly shown elsewhere (e.g., De Datta and Buresh, 1989; Schnier, 1995), specific conditions in wetland rice production often result in rather lower fertilizer N recovery than that found in upland crop production, particularly when urea, ammonium sulfate or ammonium bicarbonate were broadcast on the soil surface or into the floodwater. Examples given in Table 1 show, however, that nitrogen recovery can be substantially improved by incorporation of N fertilizer, like banding of urea or deep-point placement of urea supergranules. Furthermore, specific formulations of urea, e.g., sulfur-coated urea, or proper timing of N application could significantly improve N recovery from urea.

In conclusion, the studies presented in Table 1 show that various N fertilizers may differ considerably with respect to yield effectiveness and N recovery in the plant/soil system. There is, however, considerable scope for improvement of the utilization of each of these N sources by proper timing and a technique of N application. This requires understanding of the transformation of various N fertilizers in soils and of the associated processes influencing N availability and the various N loss pathways.

TRANSFORMATION OF NITROGEN FERTILIZERS IN SOIL AND ASSOCIATED PROCESSES CONCERNING N AVAILABILITY AND ENVIRONMENTAL POLLUTION

Plants may take up fertilizer-derived nitrogen in the form of NO_3^-, NH_4^+, urea and amino acids. However, following application, urea

TABLE 1. Crop yield, nitrogen uptake and fertilizer-N recovery as influenced by the fertilizer type, technique and time of N application

Crop	Location and soil	N form and N rate [kg ha⁻¹]	Technique and time of N application	Dry matter yield[a] [t ha⁻¹]	Shoot N uptake [kg ha⁻¹]	Apparent N recovery[b] [%]	15N recovery in plant [%]	15N recovery in soil [%]	15N loss [%]	Reference
Maize lysim.	Minn., USA sandy loam	U 180 U+NI 180	BC, S, P BC, S, P	8.4 8.4	163 171	51 55	44 41	37 50	19 9	Walters and Malzer (1990a)
Maize conv.	Iowa, USA loam	UAN 224 UAN 224	BC, S, fall BaP, I, spring	5.5 6.2	125 151	– –	13 36	18 22	69 42	Timmons and Cruse (1990)
Maize conv.	Iowa, USA 2 sites	AA 112 AA+NI 112	BaP, I, BP BaP, I, BP	5.2 5.1	124 128	54 58	39 44	– –	55 54	Blackmer and Sanchez (1988)[c]
Maize conv.	Nebr., USA silty clay loam	AS 112 AS 112	BC, I, P BC, I, AP	10.6 11.1	231 229	58 57	61 66	12 27	27 7	Bigeriego, Hauck, and Olson (1979)
Maize conv.	N.Car., USA sand	U 112 U+NI 112 NaN 112	BC, I, BP BC, I, BP BC, I, BP	3.0 6.1 2.0	47 87 32	17 53 4	– – –	– – –	– – –	Chancy and Kamprath (1982)
Maize no-till	Penn., USA silt loam	U 101 UAN 101 AS 101 AN 101	BC, S, BP BC, S, BP BC, S, BP BC, S, BP	8.0 7.7 9.0 9.0	115 106 133 138	52 43 70 75	– – – –	– – – –	– – – –	Fox and Hoffman (1981)
Maize no-till	Penn., USA silt loam	U 134 UAN 134 AN 134	BC, S, AP BC, S, AP BC, S, AP	8.2 8.9 9.2	139 151 165	44 53 63	– – –	– – –	– – –	Fox, Piekielek, and Macneal (1996)

Crop	Location, soil	N source	Rate	Method							Reference
Wheat	Mont., USA clay loam	U	101	BC, S, AP	3.8	85	41	33	–	–	Christensen and Meints (1982)
		AN	101	BC, S, AP	3.6	82	38	31	–	–	
Wheat	Pakistan calcareous silt loam	U	120	BC, I, P/S, AP	5.3	130	77	62	–	–	Hamid and Ahmad (1995)
		AS	120	BC, I, P/S, AP	5.3	131	78	62	–	–	
		AN	120	BC, I, P/S, AP	5.7	149	93	69	–	–	
Ryegrass	N. Ireland clay loam	U	100	BC, S	4.6	–	67	32	22	46	Watson et al. (1994)
		U+UI	100	BC, S	5.0	–	81	38	23	39	
		AN	100	BC, S	4.8	–	83	–	–	–	
Brome-grass	Alberta, Canada clay	U	100	BC, S	–	–	–	41	13	46	Malhi (1995)
		U	100	BaP, I	–	–	–	49	15	36	
		AN	100	BC, S	–	–	–	51	12	37	
		AN	100	BaP, I	–	–	–	51	12	37	
Rice flooded	Philippines clay	U	87	BC, S, AP	5.4	–	23	22	16	62	Craswell et al. (1985)
		USG	87	PP, I, P	6.2	–	75	47	20	33	
		SCU	87	BC, I, P	5.8	–	46	34	9	57	
		AS	87	BC, S, AP	5.4	–	32	22	12	66	
Rice flooded	Philippines	U	87	BC, I, P/BC, S, AP	7.6	105	66	41	24	35	Schnier et al. (1990)
		U	87	BaP, I, AP/BC, S, AP	8.0	108	69	70	22	8	
		USG	87	PP, I, AP	7.8	112	74	75	21	4	
Rice flooded	China sandy loam	U	90	BC, I, BP	–	–	–	28	12	60	Zhu et al. (1989)
		ABC	90	BC, I, BP	–	–	–	21	11	68	

a Grain yield of cereals and total shoot dry matter of grasses.
b Apparent N recovery = (N uptake$_F$ – N uptake$_C$)/fertilizer N applied * 100; F = fertilized plot, C = unfertilized control.
c 15N loss calculated on the basis of 15N uptake of grains and 15N recovery in soil 1 year after 15N application (Sanchez and Blackmer, 1988).

and ammonium fertilizers are subjected to rapid conversion in most arable soils (Figure 1). When urea or urea-containing fertilizers are applied, urea is normally rapidly hydrolyzed by the enzyme urease to form ammonium carbonate. $(NH_4)_2CO_3$ decomposes to produce NH_3/ NH_4^+ and one or more inorganic carbon species. The associated rise of soil pH and the proportion of NH_3 to NH_4^+ depends on the initial soil pH, but also on the pH buffering capacity of the soil (Ferguson et al., 1984). In many studies an increase in soil or floodwater pH from approximately 6-7 to about 8-9 or greater has been observed as a consequence of urea hydrolysis (e.g., Chauhan and Mishra, 1989; Clay, Malzer, and Anderson, 1990a). Shifts of the soil pH similar to those arising from the application of urea also occur following the application of ammonium bicarbonate, anhydrous ammonia and diammonium phosphate. In contrast, hydrolysis of monoammonium phos-

FIGURE 1. Transformation of nitrogen fertilizers in soil and nitrogen loss pathways.

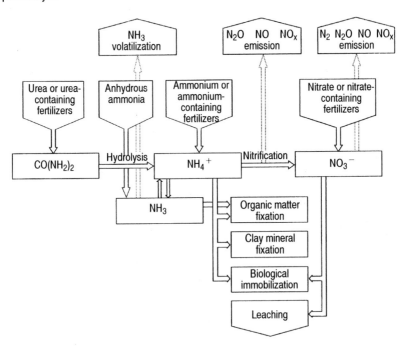

phate results in a decrease of the soil pH (Whitehead and Raistrick, 1990; Fan and MacKenzie, 1993).

Ammonium, derived from either ammonium-forming or ammonium-containing fertilizers undergoes microbial oxidation to NO_2^- and then to NO_3^-. Commonly, nitrification takes place in all arable soils where NH_4^+ is present and soil aeration is ensured. The nitrification rate is reduced in cold and acid soils; however, temperature and pH optima may vary widely among soils. Maximum nitrification rates have been found between 20 and 40°C and within a pH range of 6 to 8 (Schmidt, 1982). The nitrification process results in the release of 2 H^+ per molecule NH_4^+ and is, therefore, associated with immediate soil acidification.

Nitrogen transformation processes in soil as outlined above have numerous consequences with respect to soil fertility and environmental pollution, such as dissolution of soil organic matter through alkaline-hydrolyzing fertilizers (Sen and Chalk, 1994), acidification of the soil in the long-term due to nitrification and nitrate leaching (Bouman et al., 1995), micro- and macronutrient availability (Darusman et al., 1991), and the capacity of soils to produce or oxidize methane (Minami, 1995; Hütsch, Russell, and Mengel, 1996). It is beyond the scope of this review to deal with these aspects in more detail. However, from an economic, as well as from an environmental, point of view an assessment of the significance of various N loss pathways, namely (i) ammonia volatilization, (ii) dinitrogen and nitrogen oxide emission, and (iii) nitrate leaching, is of primary interest when comparing various N fertilizers. Furthermore, the effects of biological N immobilization and NH_4^+ fixation on N availability and, finally, the impact of NH_4^+ vs. NO_3^- uptake on crop growth and yield have to be considered. These aspects will be discussed in the following sections.

Ammonia Volatilization

Increased use of animal manure and manufactured N fertilizers have led to an increase in NH_3 volatilization from the land surface during the past decades. In addition to the direct economic loss to the farmer, redeposition of NH_4^+ may adversely affect forest and natural ecosystems, e.g., by contributing to soil acidification, forest decline (for review see Zöttl, 1990) and changes in plant species composition of natural ecosystems (e.g., Bobbink, Heil, and Raessen, 1992). Environmental factors that affect NH_3 volatilization have been discussed in numerous reviews (e.g., Terman, 1979; Fenn and Hossner, 1985). In brief, NH_3 losses increase with higher soil pH, temperature and windspeed, whereas high soil cation exchange capacity (CEC) and a high atmospheric NH_3 concentration may reduce volatilization. Ammonia volatilization from ammonium salts and urea is affected by carbonate reactions leading to the formation of $(NH_4)_2CO_3$, which is unstable and breaks down into NH_3 and CO_2. Formation of $(NH_4)_2CO_3$ from ammonium salts mainly occurs in alkaline soils and is highly dependent on the anion compound (X) of the ammonium source (Fenn and Hossner, 1985). If, for example, X is SO_4^{2-} or HPO_4^{2-}, then CaX precipitates releasing carbonate ions. The carbonate ions may react with NH_4^+ to produce $(NH_4)_2CO_3$. If X is NO_3^- or Cl^-, soluble CaX will be formed which limits the reaction with NH_4^+ to form $(NH_4)_2CO_3$. Hydrolysis of urea generally leads to the formation of $(NH_4)_2CO_3$, which makes this N source susceptible to NH_3 volatilization on both alkaline and acid soils.

Results of early studies dealing with NH_3 losses from different N sources have been summarized in a comprehensive review by Terman (1979). Results of a more recent column study in which Whitehead and Raistrick (1990) compared NH_3 volatilization from five N fertilizers applied to the surface of five contrasting soils are shown in Table 2.

TABLE 2. Cumulative NH_3 volatilization (% of N applied) from soil columns in a controlled environment 8 days after application of urea, ammonium sulfate (AS), monoammonium phosphate (MAP), diammonium phosphate (DAP) and ammonium nitrate (AN) to the surface of five soils. Data in parentheses: soil pH 24 h after incorporation of N fertilizers[a]

Soil	Soil properties			NH_3-N loss [% of N applied] (soil pH)				
	pH	$CaCO_3$ [g 100 g^{-1}]	CEC [mmol 100 g^{-1}]	Urea	AS	MAP	DAP	AN[b]
1	3.7	0.0	19.4	<1 (4.0)	<1 (3.7)	<1 (3.7)	1 (4.6)	<1 (3.5)
2	5.5	0.0	12.8	24 (7.5)	1 (5.8)	<1 (4.9)	5 (6.8)	<1 (5.5)
3	6.1	0.6	7.4	38 (7.9)	4 (6.1)	<1 (4.8)	9 (7.0)	<1 (5.9)
4	7.1	1.8	15.6	27 (8.0)	32 (7.8)	<1 (5.9)	10 (6.9)	9 (7.5)
5	7.4	75.0	11.6	43 (8.2)	49 (7.7)	35 (7.6)	53 (8.1)	8 (7.4)

[a] Source: Whitehead and Raistrick (1990).
[b] <1-18% of applied NH_4^+-N.

Ammonia volatilization was closely related to the soil pH as measured after incubation of the soils containing different N sources. Application of urea as an alkaline-hydrolyzing fertilizer resulted in large N losses from each soil even when the initial soil pH was low, with the exception of the very acid soil 1. A similar, but less pronounced, loss pattern was observed for diammonium phosphate. In clear contrast, NH_3 losses after the application of monoammonium phosphate as an acidic-hydrolyzing fertilizer were negligible in each soil, with the exception of the calcareous soil 5. Very large losses from monoammonium phosphate and diammonium phosphate on soil 5 can be explained by the formation of stable Ca phosphates which promote the formation of $(NH_4)_2CO_3$. With ammonium sulfate, NH_3 losses were negligible from the non-calcareous soils 1 to 3 but exceeded those from urea on soils 4 and 5, presumably due to the precipitation of $CaSO_4$ followed by the formation of $(NH_4)_2CO_3$. In contrast, volatilization was much less from ammonium nitrate, reflecting the fact that the application of ammonium nitrate is not associated with the production of insoluble precipitates and $(NH_4)_2CO_3$.

Differences between N sources in their susceptibility to NH_3 volatilization, as found in the column study by Whitehead and Raistrick (1990), were confirmed in numerous field experiments. Some of these studies are summarized in Table 3. Averaged over years or sites in experiments with upland crops, losses (as a percentage of applied N) varied from 3-36% for urea, 9-16% for urea-ammonium nitrate solution, 1-43% for ammonium sulfate and 0-7% for ammonium nitrate. As expected, greatest losses were found after surface application of urea, even on soils with relatively low initial pH. In contrast to soil application, relatively little NH_3 volatilized after foliar application of urea (Smith et al., 1991), presumably because of rapid absorption of most of the urea before hydrolysis took place (Gooding and Davies, 1992). Although lower than from urea, considerable losses were also measured after surface application of urea-ammonium nitrate solution (Bundy and Oberle, 1988; Fox, Piekielek, and Macneal, 1996) and diammonium phosphate (Sommer and Jensen, 1994). By contrast, NH_3 losses from ammonium nitrate were negligible in most experiments. Finally, experiments including ammonium sulfate clearly reflect the influence of soil properties on volatilization. Losses were generally small from non-calcareous soils (Freney et al., 1992; Sommer and Jensen, 1994) but large from calcareous soils (Hargrove,

TABLE 3. Ammonia volatilization (% of applied N) as affected by N fertilizer type and N management practices

Kind & location of study Soil properties Crop & period of measurement	N form & N rate [kg N ha⁻¹]		Method of application	NH₃ loss [%]	Reference
Field exp., Penn., USA Silt loam Maize, 14-25 d	U UAN	134 134	BC, S BC, S	30 16	Fox, Piekielek, and Macneal (1996)
Field exp., Australia pH 4.5-4.8 Sugar cane (trash covered)	U AS	160 160	BC, S BC, S	29 1	Freney et al. (1992)
Field exp., Australia Clay loam, pH 5.3 Wheat, 8 d	U U AN AN	100 100 100 100	BC, S BaP, I BC, S BaP, I	36 7 4 0	Bacon, Hoult, and McGarity (1986)
Field exp., Denmark Sandy loam, pH 6.1 Wheat, 18-20 d	U DAP AS AN	100 100 100 100	BC, S BC, S BC, S BC, S	25 14 <5 <2	Sommer and Jensen (1994)
Field exp., Australia Wheat, 12 d after heading	U	50	Leaf applic.	4	Smith et al. (1991)
Field exp., Texas, USA Calcareous clay Bermudagrass	AS AN	140 140	BC, S BC, S	43 7	Hargrove, Kissel, and Fenn (1977)
Field exp., Wisc., USA Silt loam, pH 6.1-7.2 Maize, 8-10 d	U UAN U-P U-Ca AN	112 112 112 112 112	BC, S BC, S BC, S BC, S BC, S	17 9 5 6 2	Bundy and Oberle (1988)
Field exp., N. Ireland Clay loam, pH 6.2 Ryegrass, 9 d	U U+nBTPT	100 100	BC, S BC, S	13 2	Watson et al. (1994)
Field exp., India Sandy loam, pH 7.3 Lemon grass, 60 d	U U U+DCD U+DCD	187 187 187 187	S I S I	16 3 36 6	Prakasa and Puttanna (1987)
Pot exp. Silt loam, pH 5.0 Flooded rice soil, 21 d	U USG SCU AS	100 100 100 100	BC, I PP, I BC, I BC, I	51 3 7 15	Vlek and Craswell (1979)

Kind & location of study Soil properties Crop & period of measurement	N form & N rate [kg N ha^{-1}]		Method of application	NH$_3$ loss [%]	Reference
Field exp., Philippines Silty clay loam, pH 5.8 Flooded rice, 9d	U U	80 80	BC, S BC, I	47 15	Fillery, Simpson, and De Datta (1986)
Field exp., Philippines 4 sites, pH 4.6-6.8 Flooded rice	U U	80 80	BC, S BC, I	10-56 10-43	Freney et al. (1990)
Field exp., India Silty clay loam, pH 7.6 Flooded rice, 10-15 d	U USG	120 120	BC, I/S PP, I	19 1	Chauhan and Mishra (1989)
Field exp., China Sandy loam, pH 8.8 Flooded rice, 51 d	U ABC	90 90	BC, I BC, I	30 39	Zhu et al. (1989)
Field exp., Thailand Clay, pH 5.1 Flood rice, 11 d	U U+nBTPT U+PPD	60 60 60	BC, S BC, S BC, S	15 5 7	Phongpan et al. (1995)

Kissel, and Fenn, 1977). In addition to the N sources listed in Table 3, NH$_3$ volatilization may also occur after the application of anhydrous ammonia. Its extent largely depends on the NH$_3$ retention capacity of the soil, the NH$_3$ concentration within the injection zone and the depth of application (Izaurralde, Kissel, and Cabrera, 1990).

Specific conditions in wetland rice production, in particular high temperatures and elevated floodwater pH resulting from depletion of CO$_2$ in the floodwater by algal growth provide an environment favoring NH$_3$ volatilization (Fillery, Roger, and De Datta, 1986). Table 3 shows that more than 50% of applied N may be volatilized when urea is broadcast to the surface of wetland rice soils or directly into the floodwater. Losses from ammonium bicarbonate may even exceed those from urea (Zhu et al., 1989). Although ammonium sulfate is also highly susceptible to NH$_3$ volatilization when applied to alkaline floodwater, losses are generally lower than those from urea. This has been explained by the fact that NH$_3$ volatilization decreases floodwater alkalinity immediately after ammonium sulfate application, whereas an increase in alkalinity due to hydrolysis of urea precedes the decrease during NH$_3$ volatilization (Vlek and Craswell, 1979).

In order to reduce NH_3 volatilization, a number of management practices have been developed. Table 3 shows that incorporation of urea in upland crop production (e.g., Bacon, Hoult, and McGarity, 1986; Prakasa and Puttanna, 1987) may substantially decrease NH_3 losses. Similar benefits have been achieved when urea was leached into the soil by natural rainfall events or supplemental irrigation following fertilizer application (Bouwmeester, Vlek, and Stumpe, 1985). In wetland rice production, incorporation of urea before flooding (Freney et al., 1990) or deep placement of urea supergranules (Vlek and Craswell, 1979; Chauhan and Mishra, 1989) has proven to be a good technique to reduce NH_3 losses.

Considerable efforts have been made to investigate the effects of urea amendments such as (i) acidic materials, (ii) neutral salts, (iii) urease inhibitors, and (iv) urea coatings. Decrease of microsite pH and thus a reduction of NH_3 losses could be achieved by use of urea phosphates (Bundy and Oberle, 1988) and also by mixing urea with triple superphosphate or monoammonium phosphate (Fan and MacKenzie, 1993). Similar effects could be achieved by use of urea-KCl and urea-$CaCl_2$ (Bundy and Oberle, 1988; Christianson et al., 1995). Studies dealing with the effectiveness of urease inhibitors in reducing NH_3 volatilization did not lead to consistent results (e.g., Beyrouty, Sommers, and Nelson, 1988). However, recent studies indicated that NH_3 volatilization may be considerably reduced by effective urease inhibitors (Watson et al., 1994; Phongpan et al., 1995). Freney et al. (1995) reported that additional use of algal inhibitors may lead to a further reduction in NH_3 volatilization in wetland rice production. In contrast, the application of ammonium-forming fertilizers together with nitrification inhibitors resulted in increased NH_3 volatilization in some studies (e.g., Prakasa and Puttanna, 1987), whereas in other studies nitrification inhibitors had no effect on volatilization (e.g., Clay, Malzer, and Anderson, 1990a). Finally, NH_3 volatilization can be reduced by the use of slow-release fertilizers such as sulfur-coated urea (Vlek and Craswell, 1979).

Dinitrogen and Nitrogen Oxide Emission

Gaseous N losses from soil occur not only through NH_3 volatilization but also via dinitrogen (N_2) and nitrogen oxide (NO, N_2O, NO_x) emissions. Among the latter, N_2O has been studied most intensively. Heterotrophic denitrification is the major source for N_2 loss from agricultural soils and contributes also to the emission of nitrogen oxides. Favored formation of N_2O relative to N_2 only occurs under conditions marginal for denitrification. However, the evolution of N_2O also occurs during nitrification (Bremner and Blackmer, 1978). This has been attributed to the use of NO_2^- by NH_4^+ oxidizers as an alternative electron acceptor when O_2 is limiting, and to chemodenitrification by which intermediates between NH_4^+ and NO_2^-, or NO_2^- itself, chemically decompose to N_2O (for review see Granli and Bøckman, 1994). From an economic point of view, losses of N_2 are undoubtedly of major importance. However, since N_2O contributes to both absorption of thermal radiation and thus to global warming and to the destruction of the ozone layer in the stratosphere, increased atmospheric N_2O concentration has led to considerable environmental concern.

The extend to which N_2O is produced and lost from the soil is influenced by a number of environmental factors, such as oxygen availability, C availability, pH, and temperature, as well as by management practices, such as fertilizer application (type, rate, timing, and technique), organic manuring and irrigation (Sahrawat and Keeney, 1986; Granli and Bøckman, 1994). All these factors cause a great variability in N_2O losses. Eichner (1990), who collected data from 104 field experiments, found that the application of mineral N fertilizers enhanced N_2O emissions from the soil by a factor between 1.1 and 15.3 compared to unfertilized control plots. A comparison of fertilizer types revealed a clear trend in that application of anhydrous ammonia gave markedly higher N_2O emissions (2.7% of applied N) than any other N source (up to 0.4% of applied N). Examples given in Table 4 confirm this observation. Strong evolution of N_2O after anhydrous ammonia application has been explained by NO_2^- accumulation caused by pH rise up to pH 10 in the injection zone, which appears to inhibit the activity of NO_2^- oxidizers more than the activity of NH_4^+ oxidizers (Peoples, Freney, and Mosier, 1995). Similar effects can be expected when other alkaline-hydrolyzing fertilizers (such as urea or diammonium phosphate) are applied in bands or nests. Table 4 further shows that N_2O emission after the application of ammonium or urea is often higher than after the application of nitrate. Each of these studies indicated that the bulk of N_2O was formed during nitrification and not during denitrification. It has been stressed, however, that N_2O emissions from nitrate fertilizers may exceed those from ammonium fertil-

TABLE 4. Nitrous oxide emission as affected by N fertilizer type

Kind & location of study Soil properties Crop & period of measurement	N form & N rate [kg N ha⁻¹]		N$_2$O emission [g N ha⁻¹] [%][a]		Reference
Pot exp.		0	9	–	Flessa et al.
Loess topsoil	ACl	100	73	0.06	(1996)
Unplanted, 50 d	KN	100	12	<0.01	
Field exp., Iowa, USA		0	400	–	Bremmer,
Fallow soil, 42 d	AA	250	11300	4.36	Breitenbeck, and Blackmer (1981)
Field exp., Australia		0	33	–	Magalhaes,
pH 8.5	AA	80	70	0.05	Chalk, and
Fallow soil, 27-29 d	AA+NP	80	38	0.01	Strong (1984)
Field exp., Iowa, USA		0	55	–	Breitenbeck,
Fallow soil, 43 d	U	250	288	0.09	Blackmer, and
	AS	250	355	0.12	Bremner (1980)
	CaN	250	73	0.01	
Field exp., NY, USA		0	300	–	Duxbury and
Silt loam	U	120	2500	1.83	McConnaughey
Maize, 83 d	CaN	120	300	0.00	(1986)
Field exp., Col., USA		0	109	–	Bronson, Mosier,
Clay loam, pH 7.2	U	218	2746	1.21	and Bishnoi
Maize, irrigated, 97 d	U+NP	218	1088	0.45	(1992)
	U+ECC	218	1251	0.52	

[a] % of applied N = {([g] N$_2$O-N from fertilized plots − [g] N$_2$O-N from unfertilized plots)/[g] fertilizer-N} × 100.

izers under conditions favoring denitrification (Granli and Bøckman, 1994). In wetland rice production, N$_2$O emissions appear to be small after ammonium sulfate or urea application under continuous flooding conditions (Freney et al., 1981; Lindau et al., 1990). However, considerable amounts of N$_2$O can be emitted when rice is direct-seeded followed by several cycles of wetting and drying (Keerthisinghe, Freney, and Mosier, 1993). There are a number of studies showing that N$_2$O losses may be substantially reduced by effective inhibition of nitrification in upland soils (Magalhaes, Chalk, and Strong, 1984; Bronson, Mosier, and Bishnoi, 1992) as well as in rice soils (Keerthisinghe, Freney, and Mosier, 1993). Total (N$_2$ + N$_2$O) losses due to denitrification may be dramatically increased from submerged soils after the application of nitrate fertilizers (Lindau et al., 1990). However, Aulakh, Doran, and Mosier (1992) emphasized that often factors other than NO_3^- supply, in particular water content and C availability, limit denitrification in arable soils. Von Rheinbaben (1990), evaluating 38 field experiments, estimated that denitrification losses may reach up to 10% or more of the fertilizer input but they are often lower and of minor importance with respect to fertilizer N recovery. On the other hand, agricultural practices such as irrigation, organic manuring or reduced/non-tillage may considerably increase denitrification.

Table 5 shows that, for example, long-term application of sewage

TABLE 5. Dinitrogen plus nitrous oxide emissions as affected by fertilizer type and N management practices

Kind & location of study Soil properties Crop & period of measurement		N form & N rate [kg N ha^{-1}]		N$_2$O + N$_2$ emission [g N ha^{-1}] [%][a]		Reference
Field exp.	Sew. Sludge					
Germany	–	AN	120	3.00	–	Benckiser et al.
Silty sand	+	AN	120	15.00	–	(1987)
240 d						
Field exp., NY, USA			0	1.90		Duxbury and
Silt loam		U	120	4.90	2.50	McConnaughey
Maize, 83 d		CaN	120	2.30	0.33	(1986)
Field exp.			0	<1.00		Jordan (1989)
N. Ireland	well drained	U	300	10.00	3	
Ryegrass	clay loam	AN	300	29.00	10	
365 d						
			0	<1.00		
	poorly	U	300	31.00	10	
	drained clay	AN	300	79.00	26	
Field exp., Philippines		U, BC, S	80	2-40	–	Freney et al.
4 sites		U, BC, I	80	12-45	–	(1990)
Flooded rice						

[a] % of applied N = {([kg] N$_2$O-N + N$_2$ from fertilized plots – [kg] N$_2$O-N + N$_2$ from unfertilized plots)/[kg] fertilizer-N} × 100

sludge increased the N losses five-fold due to denitrification from a silty sand treated with ammonium nitrate (Benckiser et al., 1987). Similarly, relatively high losses can be expected from permanent grassland with high organic C content combined with high fertilizer application rates, particularly from poorly drained soils (Jordan, 1989). Results are not consistent with respect to different N sources. This may be caused by different environmental conditions, being either unfavorable or favorable for denitrification. In the former case, $N_2 + N_2O$ losses from ammonium-based fertilizers may exceed those from nitrate fertilizers (Duxbury and McConnaughey, 1986, see also Table 4), whereas in the latter case they are considerably larger from nitrate fertilizers (Jordan, 1989).

Flooded rice soils present a favorable environment for denitrification, provided that NO_3^- is present. Although N fertilizer inputs to wetland rice are almost exclusively in the form of urea or ammonium salts and nitrification is controlled by predominantly anoxic conditions in these soils, significant amounts of NH_4^+ may be nitrified in the oxidized rhizosphere of rice roots and within a thin oxidized soil surface layer. When this nitrate moves into adjacent reduced layers, rapid denitrification occurs (Savant and De Datta, 1982). Measurements of gaseous N losses due to denitrification from rice fields led to quite variable results. Very low losses as often found using the ^{15}N chamber technique were partly explained by methodological problems associated with this direct approach (for review see Buresh and De Datta, 1990). As an alternative, denitrification losses have been estimated indirectly as the difference between total ^{15}N loss in a ^{15}N balance and measured NH_3 volatilization loss. Using this approach, Freney et al. (1990) found that NH_3 volatilization (10-56% of applied N) and estimated denitrification losses (3-56% of applied N) varied greatly between experimental sites and between N management practices. Denitrification losses were generally low where NH_3 volatilization was high and vice versa. Incorporating applied urea into the flooded soil was suitable to reduce N losses by NH_3 volatilization but increased losses by denitrification. The authors concluded that the efficiency of fertilizer N will be improved only when both N loss pathways, NH_3 volatilization and denitrification, are controlled simultaneously. This might be achieved by incorporation of urea in combination with the use of nitrification inhibitors (Keerthisinghe et al., 1996). Similarly, the application of urea together with nitrification inhibitors resulted in higher N recoveries in upland crop production under conditions favoring denitrification, thus providing indirect evidence that effective inhibition of nitrification may reduce gaseous N losses due to denitrification (Freney et al., 1993).

Nitrate Leaching

Leaching of nitrogen is a major loss pathway from agricultural soils under high water surplus conditions, particularly in soils with low water holding capacity and with shallow rooting crops. Leaching of N occurs primarily in the nitrate form since (i) urea, although mobile in soil, is rapidly converted to NH_4^+, and (ii) NH_4^+ is adsorbed to the soil cation exchange complex. The nitrification of ammonium is therefore a key step in the chain of events leading to N losses through leaching. Besides other N management practices to reduce nitrate leaching, such as accurate prediction of crop N requirements and proper timing of N fertilizer applications, experimental efforts have been focused on amendment of ammonium or ammonium-forming fertilizers with chemicals to delay nitrification.

Some examples of the effectiveness of nitrification inhibitors as found in column and lysimeter studies are shown in Table 6. Using unplanted and periodically leached soil columns, Timmons (1984) found that the nitrate concentration in the percolate and total leaching losses could be moderately reduced by amending urea with nitrapyrin. Similarly, nitrapyrin reduced leaching losses from field lysimeters planted with maize. Benefits were greatest when large amounts of fertilizer N were applied and high precipitation favored leaching (Owens, 1987). With lower rates of fertilizer N applied and less water surplus, differences between nitrapyrin-treated and untreated lysimeters were small but detectable (Timmons, 1984; Walters and Malzer, 1990a, b). Similarly to manufactured fertilizers, leaching losses of nitrate during the vegetation-free winter period could be reduced by amending animal slurry with dicyandiamide (Amberger, Gutser, and Vilsmeier, 1982).

In field experiments, the impact of different N fertilizer types or nitrification inhibitors on nitrate leaching has often been derived indirectly from yield responses and fertilizer N recoveries in plants and soil. Although this approach provides no reliable information with regard to various loss pathways, there are some studies showing that amending autumn-applied ammonium/urea fertilizers with nitrifica-

TABLE 6. Nitrogen losses through leaching after application of urea with or without nitrapyrin in soil column and lysimeter studies

Kind & duration of study	Location, soil & crop	N form & N rate [kg ha⁻¹]		Precipitation + irrigation [L m⁻²]ᵃ	Soil water percolation [L m⁻²]ᵃ	Average N concentration in percolate [mg L⁻¹]ᵃ	Leaching loss [kg N ha⁻¹]ᵃ	Reference
Soil columns (d: 1.2 m)ᵇ 105 days	Lab sandy loam unplanted	U U+NP	0 224 224	572 572 572	555 554 556	18.2 53.2 57.6	102 296 266	Timmons (1984)
Field lysimeter (d: 1.2 m) 3 years	Minnesota, USA sandy loam maize	U U+NP	224 224	– –	263 261	61.6 57.4	161 149	Timmons (1984)
Field lysimeter (d: 2.4 m) 6 years	Ohio, USA silt loam	U U+NP	336 336	1186 1186	513 502	31.2 23.3	160 117	Owens (1987)
Field lysimeter (d: 1.2 m) 2 years	Nebraska, USA sandy loam maize	U U+NP	0 180 180	623 623 623	147 122 131	7.5 31.1 26.0	11 38 34	Walters and Malzer (1990a, b)

ᵃ Average per year in long-term field experiments.
ᵇ Depth of soil column/lysimeter.

tion inhibitors improved grain yield of winter wheat (Huber et al., 1980) and decreased downward movement of fertilizer ^{15}N in the soil during winter (Aulakh and Rennie, 1984; Bronson et al., 1991). Similarly, in a field experiment reported by Frye (1977), soil analysis indicated that nitrate leaching into deeper soil layers was reduced and grain yield was increased subsequent to fall-applied slow release sulfur-coated urea fertilizer compared to sodium nitrate or urea. However, N application in spring generally resulted in improved grain yields compared with fall N application, irrespective of the N fertilizer type. This result clearly indicates that N fertilizer application in spring is the more suitable N management strategy from an environmental point of view, particularly in subhumid/humid regions.

But even for spring application, large differences in the effectiveness of various N fertilizers were found, provided soil and weather conditions during the growing season favored nitrate leaching. For example, in a 2-years field study conducted on a sandy soil in N. Carolina, Chancy and Kamprath (1982) observed little effect of N fertilizer type on maize grain yield in a year with low rainfall after N application (Figure 2). However, in the second year, when high rainfall following N application favored nitrate leaching, corn grain yield and N recovery were dramatically reduced in the urea and particularly in the sodium nitrate treatments compared to the urea plus nitrapyrin

FIGURE 2. Grain yield of maize and apparent recovery of fertilizer nitrogen as affected by N fertilizer type (U = urea, NaN = sodium nitrate), application of a nitrification inhibitor (NI), and weather conditions after N application. Adapted from Chancy and Kamprath (1982).

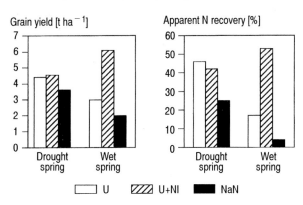

treatment. There are various other studies showing beneficial effects when nitrification inhibitors along with urea/ammonium fertilizers or slow release fertilizers were applied in spring under environmental conditions that were conducive to leaching, particularly on light-textured soils in combination with high rainfall (e.g., Malzer et al., 1989; Francis, Doran, and Lohry, 1993). On the other hand, the use of stabilized ammonium fertilizers had often no beneficial effect on yield and was therefore cost-ineffective on fine-textured soils, even under high rainfall conditions (Blackmer and Sanchez, 1988; Malzer et al., 1989; Buerkert, Horlacher, and Marschner, 1995). Finally, it should be emphasized that delayed N application closer to the time of maximum crop uptake may be at least as effective in reducing nitrate leaching losses as the use of nitrification inhibitors (Malzer et al., 1989).

Immobilization of Fertilizer Nitrogen and Mineralization of Soil Nitrogen

The availability of fertilizer nitrogen is affected by its involvement in the mineralization-immobilization turnover in soil. The net effects of this turnover, net mineralization or net immobilization, are easy to estimate by plant and soil analysis. For example, Engels and Kuhlmann (1993), who conducted a series of field experiments with cereals in Germany, observed a shift from net mineralization to net immobilization with increasing N fertilizer rate. However, to examine the two opposing processes (mineralization and immobilization) separately, the use of ^{15}N-labeled fertilizers is required.

With respect to biological immobilization, comparisons of various N sources often revealed that both nitrate and ammonium are good substrates for assimilation by the soil microflora. Jansson, Hallam, and Bartholomew (1955), however, were the first who showed that microorganisms utilized ammonium in preference over nitrate when both ions were supplied in sufficient and equal quantities. This observation was confirmed in numerous incubation studies (e.g., Recous and Mary, 1990; Azam, Simmons, and Mulvaney, 1993) and also in field experiments. For example, Aulakh and Rennie (1984) found that N immobilization from fall-applied urea (24-26% of applied N) exceeded that from potassium nitrate (15-21% of applied N). Increased N immobilization with ammonium fertilizers may be still detectable several years after N application (Riga, Fischer, and Van Praag, 1980). Since retardation of nitrification increases the persistence of NH_4^+ in soils, the application of ammonium-forming fertilizers together with the nitrification inhibitor has often resulted in increased biological immobilization in incubation studies (Guiraud, Marol, and Fardeau, 1992) and also in field experiments (Juma and Paul, 1983; Bronson et al., 1991). Immobilization may be considerably reduced by banding or nesting of ammonium-forming fertilizers instead of broadcasting (Yadvinder-Singh et al., 1994), particularly when crop residues with a high C:N ratio are present in the soil (Tomar and Soper, 1981).

Immobilization of inorganic N into the organic N pool is accompanied by a simultaneous mineralization of organic N into inorganic forms. It has been often shown that gross mineralization of soil N increased following the application of fertilizer N in incubation studies without plants (Broadbent, 1965) and also in field studies with plants (Riga, Fischer, and van Praag, 1980; Hamid and Ahmad, 1993). When different N sources were compared, gross mineralization was often greater after the application of ammonium fertilizers than after the application of nitrate fertilizers (Azam, Simmons, and Mulvaney, 1993; Azam, Mulvaney, and Simmons, 1995). This stimulation of soil N mineralization by fertilizer N and resulting enhanced uptake of soil N by crops has been termed the "priming" effect or "added nitrogen interaction" (ANI = N derived from fertilized soil-N derived from unfertilized soil). Jenkinson, Fox, and Rayner (1985), who comprehensively discussed possible reasons of this phenomenon, emphasized that ANIs can be "real" (caused, for example, by salt and pH effects of the fertilizer on microorganism turnover in soil, or by an increase of the volume of soil explored by the roots) but they are more frequently "apparent," caused by pool substitution or isotope displacement reactions. In this light, studies showing increased ANIs after application of ammonium fertilizers compared to nitrate fertilizers have to be interpreted carefully. This may be illustrated by an incubation study of Wickramasinghe, Rodgers, and Jenkinson (1985), who examined the effects of ^{15}N-labeled urea, ammonium sulfate and potassium nitrate on immobilization of fertilizer N and mineralization of soil N in an acid tropical soil and a neutral temperate soil (Table 7). In accordance with the studies cited above, application of urea and ammonium sulfate resulted in both increased gross immobilization and increased total ANI compared to the application of potassium nitrate. However, net mineralization tended to be greater with potassium nitrate and therefore "real" ANI was larger when potassium nitrate, instead of urea or ammonium sulfate, was applied.

TABLE 7. Nitrogen transformations in acid tropical soil and a neutral temperate soil after 40 days incubation with and without N fertilizers urea, ammonium sulfate (AS) and potassium nitrate (KN); ANI = added nitrogen interaction, SE = standard error of treatment means[a]

	Net N mineralization	Gross immo-bilization[b]	Total ANI[c]	Real ANI[d]	Apparent ANI[e]
			[μg N g^{-1} dry soil]		
Acid tropical soil					
Control	7.4	–	–	–	–
Urea	9.2	11.8	15.9	1.8	14.1
AS	8.6	11.6	14.4	1.2	13.2
KN	14.1	1.8	7.5	6.7	0.8
SE	1.7	<0.1	0.8	2.0	–
Neutral temperate soil					
Control	42.0	–	–	–	–
Urea	50.9	16.0	27.9	8.9	19.9
AS	54.6	14.8	29.9	12.6	17.3
KN	55.3	2.8	18.8	13.3	5.5
SE	1.6	<0.1	3.0	2.0	–

[a] From Wickramasinghe, Rodgers, and Jenkinson (1985), with kind permission from Elsevier Science Ltd, Kidlington, UK.
[b] Fertilizer-derived organic ^{15}N.
[c] (Unlabeled NO_3-N + NH_4-N in fertilized soil after 40 d) – (NO_3-N + NH_4-N in control after 40 d).
[d] (Net N mineralized by fertilized soil in 40 d) – (Net N mineralized by control in 40 d).
[e] (Total ANI) × (Real ANI).

The involvement of fertilizer N in the mineralization-immobilization turnover in soil may, undoubtedly, affect N uptake and crop yield. For example, reduced grain yields of no-tillage corn compared to conventional tillage corn could be explained by an increased N immobilization in the no-tillage system (Kitur et al., 1984). With respect to different N sources, there are indications that increased immobilization of ammonium may reduce grain yield when the N supply is low and soil microbes compete effectively for ammonium (Bronson et al., 1991; Crawford and Chalk, 1993). However, under conditions of adequate N supply, N uptake and crop yield were not affected by increased gross immobilization of ammonium in many studies (Juma and Paul, 1983; Bronson et al., 1991; Crawford and Chalk, 1993).

Ammonia and Ammonium Fixation

In addition to biological immobilization, the availability of added N may be reduced by chemical fixation of NH_3 directly into the organic matter of the soil and by fixation of NH_4^+ into clay minerals (Nommik and Vahtras, 1982). Since a rise in soil pH increases the polarization of the reactive groups on organic matter, fixation of NH_3 on organic matter is stimulated following the application of alkaline-hydrolyzing fertilizers such as urea, or after the application of anhydrous ammonia. Inhibition of nitrification may increase this effect by maintaining a high soil pH (Clay, Malzer, and Anderson, 1990b). There is, however, little experimental evidence that organic matter fixation contributes substantially to differences in the efficacy of various fertilizer types in mineral soils under field conditions.

Fixation of NH_4^+ into clay minerals depends highly on the presence of 2:1 clay minerals and the potassium status of the soil (Nommik and Vahtras, 1982). Comparing different soils, Green, Blackmer, and Yang (1994) recovered between 1 and 23% of labeled urea-N as fixed NH_4^+ after a 20-day incubation period. In a field experiment, N recovery of N fertilizer as fixed NH_4^+ at the end of the growing season of maize was 5% in a soil with relatively low clay content but 18% in a soil with greater clay content, particularly of vermiculite and mica (Liang and MacKenzie, 1994). Release of fixed NH_4^+ in the second year was smaller from the soil with high fixation capacity, indicating that NH_4^+ fixation in this soil provided a more permanent sink. When comparing different N sources, Scherer and Mengel (1986) found that ammonium sulfate was at least as good an N source for *Lolium multiflorum* as calcium nitrate in a weakly NH_4^+-fixing soil. However, yield and apparent N recovery were significantly reduced in the ammonium sulfate treatment on soils which strongly fixed added NH_4^+. Thus, on certain soil types, NH_4^+ fixation may contribute to low N recovery from ammonium fertilizers. However, the agronomic significance of NH_4^+ fixation should not be over-emphasized since the great majority of arable surface soils fix much less NH_4^+ than found in many studies with selected soils (Nommik and Vahtras, 1982). When different soil N fractions were compared, organic N contributed much more to fertilizer-derived soil N than fixed NH_4^+ in many soils (Juma and Paul, 1983; Wickramasinghe, Rodgers, and Jenkinson, 1985).

GROWTH AND YIELD OF ARABLE CROPS
AS AFFECTED BY AMMONIUM vs. NITRATE NUTRITION

When discussing the efficacy of N fertilizers, it should be emphasized that the application of various N sources may not only affect the N availability in soils but also numerous processes in the plant/soil system and within the plant. The processes in the plant/soil system may influence the chemical availability of mineral elements due to root-induced pH changes in the rhizosphere (Thomson, Marschner, and Römheld, 1993) and also the spatial availability of mineral elements and water due to morphological modifications of the root system (Bloom, Jackson, and Smart, 1993). Once taken up, nitrate and ammonium have contrasting effects on many physiological processes in the plant (for review see Gerendás et al., 1997; Wiesler, 1997), which in turn may affect growth and yield of plants. Based on numerous nutrient solution studies showing that many plant species grow best when a mixture of both ammonium and nitrate is supplied (e.g., Cox and Reisenauer, 1973), considerable effort has been made to study the effect of an "enhanced ammonium supply" on reproductive yield of arable crops (Bock, 1986). Table 8 summarizes data obtained from the literature with respect to relative yield and yield components of wheat and maize fertilized with a mixed nitrogen source (up to 50% ammonium) as compared to sole nitrate supply. Mixed nitrogen nutrition generally increased grain yield of wheat and maize when plants were grown in nutrient solution or in gravel hydroponic systems. With regard to yield structure of wheat, mixed nitrogen nutrition generally increased the number of ears per plant. Results are not consistent with respect to the number of grains per ear, and thousand-grain weight was hardly influenced by the nitrogen form. Increased grain yield of maize was the result of more kernels per plant, whereas thousand-kernel weight was less and not consistently affected by the form of nitrogen supply.

Table 8 also includes data from field experiments where soil analyses (Pan et al., 1984; Smiciklas and Below, 1992), xylem sap analyses (Buerkert, Horlacher, and Marschner, 1995) or daily fertigation with different N sources (Yibirin, Johnson, and Eckert, 1996) indicated that the application of urea/ammonium fertilizers amended with nitrification inhibitors actually resulted in an enhanced NH_4^+ nutrition. In these experiments enhanced NH_4^+ nutrition resulted in both moderately increased and moderately decreased grain yields. Obviously, the

TABLE 8. Relative yield and yield structure of spring wheat and maize due to an enhanced ammonium supply (31-50% ammonium nitrogen in nutrients solution experiments, ammonium or urea supply plus NI in field experiments) compared to N supply predominantly in the nitrate form (96-100% nitrate nitrogen in nutrient solution experiments, nitrate supply in field experiments)

Culture system	Yield [%]	Number of ears per plant [%]	Number of grains per ear/plant[a] [%]	Thousand-grain weight [%]	Reference
Wheat					
Nutrient solution	128-167	143-153	90-104	100-104	Heberer and Below (1989)[b]
Nutrient solution	137	121	113	99	Wiesler (1997)
Pot experiment	121	158	78	–	Leyshon, Campbell, and Warder (1980)
Maize					
Gravel hydroponics	112-117	–	109	106	Alexander, Miller, and Beauchamp (1991)[b]
Gravel hydroponics	107-114	–	104-119	94-102	Gentry and Below (1993)[b]
Field experiment	91-120	97-109	–	–	Pan et al. (1984)[b]
Field experiment	101-108	100-117	97-108	101-106	Smiciklas and Below (1992)[b]
Field experiment	90-98	–	–	–	Buerkert, Horlacher, and Marschner (1995)[c]
Field experiment	97-109	–	–	–	Yibirin, Johnson, and Eckert (1996)[c]

[a] Number of grains per ear with wheat and number of grains per plant with maize.
[b] Ranges refer to comparisons of genotypes.
[c] Ranges refer to comparisons of years.

relative superiority of either form of nitrogen was highly dependent on the cultivar used and on environmental conditions which may vary from year to year. Since the availability of ammonium in soils is not easily controlled over a longer time period and the various yield components respond differently to an enhanced ammonium supply, additional research is needed to identify "critical periods" of plant development in which the determination of certain yield components might be improved by an enhanced ammonium supply. Finally, long-term effects of an enhanced ammonium supply on soil fertility should be considered.

CONCLUSIONS

To meet the food requirements of a growing world population, global N fertilizer consumption will need to increase in the future (Bumb, 1995). However, more efficient use of N fertilizers will be essential for improving the economic output of the farm and reducing the risk of environmental pollution. The main pathways by which fertilizer N may be lost from the plant/soil system, i.e., NH_3 volatilization, denitrification and nitrate leaching are well recognized. Undoubtedly, the use of N fertilizer types that are least susceptible to loss under certain environmental conditions would provide a useful strategy to improve N fertilizer efficiency. Large differences between urea or ammonium sulfate and ammonium nitrate in their susceptibility to NH_3 volatilization on calcareous soils are an impressive example. However, just the example of urea shows that there is also considerable scope to improve the efficiency of each N source by proper timing and technique of N application, even under conditions favoring N losses. Use of the enormous research outcome in this area highly depends on the policy environment as well as the acceptance by the farmers. Further research is needed to substantiate yield improvements by an "enhanced ammonium supply" under field conditions.

REFERENCES

Alexander, K.G., M.H. Miller, and E.G. Beauchamp. (1991). The effect of an NH_4^+-enhanced nitrogen source on the growth and yield of hydroponically grown maize (*Zea mays* L.). *Journal of Plant Nutrition* 14: 31-44.

Amberger, A., R. Gutser, and K. Vilsmeier. (1982). N-Wirkung von Rindergülle unter Zusatz von Dicyandiamid bzw. Stroh in Gefäß- und Lysimeterversuchen. *Zeitschrift für Pflanzenernährung und Bodenkunde* 145: 337-346.

Aulakh, M.S. and D.A. Rennie. (1984). Transformations of fall-applied nitrogen-15-labelled fertilizers. *Soil Science Society of America Journal* 48: 1184-1189.

Aulakh, M.S., J.W. Doran, and A.R. Mosier. (1992). Soil denitrification–significance, measurement, and effects of management. *Advances in Soil Science* 18: 1-57.

Azam, F., F.W. Simmons, and R.L. Mulvaney. (1993). Immobilization of ammonium and nitrate and their interaction with native N in three Illinois Mollisols. *Biology and Fertility of Soils* 15: 50-54.

Azam, F., R.L. Mulvaney, and F.W. Simmons. (1995). Effects of ammonium and nitrate on mineralization of nitrogen from leguminous residues. *Biology and Fertility of Soils* 20: 49-52.

Bacon, P.E., E.H. Hoult, and J.W. McGarity. (1986). Ammonia volatilization from fertilizers applied to irrigated wheat soils. *Fertilizer Research* 10: 27-42.

Benckiser, G., G. Gaus, K.-M. Syring, and D. Sauerbeck. (1987). Denitrification losses from an Inceptisol field treated with mineral fertilizer or sewage sludge. *Zeitschrift für Pflanzenernährung und Bodenkunde* 150: 241-248.

Beyrouty, C.A., L.E. Sommers, and D.W. Nelson. (1988). Ammonia volatilization from surface-applied urea as affected by several phosphoroamide compounds. *Soil Science Society of America Journal* 52: 1173-1178.

Bigeriego, M., R.D. Hauck, and R.A. Olson. (1979). Uptake, translocation and utilization of ^{15}N-depleted fertilizer in irrigated corn. *Soil Science Society of America Journal* 43: 528-533.

Blackmer, A.M. and C.A. Sanchez. (1988). Response of corn to nitrogen-15-labeled anhydrous ammonia with and without nitrapyrin in Iowa. *Agronomy Journal* 80: 95-102.

Bloom, A.J., L.E. Jackson, and D.R. Smart. (1993). Root growth as a function of ammonium and nitrate in the root zone. *Plant, Cell, and Environment* 16: 199-206.

Bobbink, R., G.W. Heil, and M.B.A.G. Raessen. (1992). Atmospheric deposition and canopy exchange processes in heathland ecosystems. *Environmental Pollution* 75: 29-37.

Bock, B.R. (1986). Increasing cereal yields with higher ammonium/nitrate ratio: Review of potentials and limitations. *Journal of Environmental Science and Health* A21: 723-758.

Bouman, O.T., D. Curtin, C.A. Campbell, V.O. Biederbeck, and H. Ukrainetz. (1995). Soil acidification from long-term use of anhydrous ammonia and urea. *Soil Science Society of America Journal* 59: 1488-1494.

Bouwmeester, R.J.B., P.L.G. Vlek, and J.M. Stumpe. (1985). Effect of environmental factors on ammonia volatilization from a urea-fertilized soil. *Soil Science Society of America Journal* 49: 376-381.

Breitenbeck, G.A., A.M. Blackmer, and J.M. Bremner. (1980). Effects of different nitrogen fertilizers on emission of nitrous oxide from soil. *Geophysical Research Letters* 7: 85-88.

Bremner, J.M. and A.M. Blackmer. (1978). Nitrous oxide: emission from soils during nitrification of fertilizer nitrogen. *Science* 199: 295-296.

Bremner, J.M., G.A. Breitenbeck, and A.M. Blackmer. (1981). Effect of anhydrous ammonia fertilization on emission of nitrous oxide from soils. *Journal of Environmental Quality* 10: 77-80.

Broadbent, F.E. (1965). Effect of fertilizer nitrogen on the release of soil nitrogen. *Soil Science Society of America Proceedings* 29: 692-696.

Bronson, K.F., J.T. Touchton, R.D. Hauck, and K.R. Kelley. (1991). Nitrogen-15 recovery in winter wheat as affected by application timing and dicyandiamide. *Soil Science Society of America Journal* 55: 130-135.

Bronson, K.F., A.R. Mosier, and S.R. Bishnoi. (1992). Nitrous oxide emissions in irrigated corn as affected by nitrification inhibitors. *Soil Science Society of America Journal* 56: 161-165.

Buerkert, B., D. Horlacher, and H. Marschner. (1995). Time course of nitrogen in soil solution and nitrogen uptake in maize plants as affected by form of application and application time of fertilizer nitrogen. *Journal of Agronomy and Crop Science* 174: 325-336.

Bumb, B.L. (1995). World nitrogen supply and demand: an overview. In *Nitrogen Fertilization in the Environment*, ed. P.E. Bacon. New York, USA: Marcel Dekker, pp. 1-40.

Bundy, L.G. and S.L. Oberle. (1988). Evaluation of methods for control of ammonia volatilization from surface-applied urea-containing fertilizers. *Journal of Fertilizer Issues* 5: 24-30.

Buresh, R.J. and S.K. De Datta. (1990). Denitrification losses from puddled rice soils in the tropics. *Biology and Fertility of Soils* 9: 1-13.

Chancy, H.F. and E.J. Kamprath. (1982). Effect of nitrapyrin on N response of corn on sandy soils. *Agronomy Journal* 74: 565-569.

Chauhan, H.S. and B. Mishra. (1989). Ammonia volatilization from a flooded rice field fertilized with amended materials. *Fertilizer Research* 19: 57-63.

Christensen, N.W. and V.W. Meints. (1982). Evaluating of N fertilizer sources and timing for winter wheat. *Agronomy Journal* 74: 840-844.

Christianson, C.B., G. Carmona, M.O. Klein, and R.G. Howard. (1995). Impact on ammonia volatilization losses of mixing KCl of high pH with urea. *Fertilizer Research* 40: 89-92.

Clay, D.E., G.L. Malzer, and J.L. Anderson. (1990a). Ammonia volatilization from urea as influenced by soil temperature, soil water content, and nitrification and hydrolysis inhibitors. *Soil Science Society of America Journal* 54: 263-266.

Clay, D.E., G.L. Malzer, and J.L. Anderson. (1990b). Tillage and dicyandiamide influence on nitrogen fertilizer immobilization, remineralization, and utilization by maize (*Zea mays* L.). *Biology and Fertility of Soils* 9: 220-225.

Cox, W.J. and H.M. Reisenauer. (1973). Growth and ion uptake by wheat supplied nitrogen as nitrate, or ammonium, or both. *Plant and Soil* 38: 363-380.

Craswell, E.T. and D.C. Godwin. (1984). The efficiency of nitrogen fertilizers applied to cereals in different climates. In *Advances in Plant Nutrition*, Volume 1, eds. P.B. Tinker and A. Läuchli. New York, USA: Praeger Publishers, pp. 1-55.

Craswell, E.T., S.K. De Datta, C.S. Weeraratne, and P.L.G. Vlek. (1985). Fate and efficiency of nitrogen fertilizers applied to wetland rice. 1. The Philippines. *Fertilizer Research* 6: 49-63.

Crawford, D.M. and P.M. Chalk. (1993). Sources of N uptake by wheat (*Triticum aestivum* L.) and N transformations in soil treated with a nitrification inhibitor (nitrapyrin). *Plant and Soil* 149: 59-72.

Darusman, L.R. Stone, D.A. Whitney, K.A. Janssen, and J.H. Long. (1991). Soil properties after twenty years of fertilization with different nitrogen sources. *Soil Science Society of America Journal* 55: 1097-1100.

De Datta, S.K. and R.J. Buresh. (1989). Integrated nitrogen management in irrigated rice. *Advances in Soil Science* 10: 143-169.

Duxbury, J.M. and P.K. McConnaughey. (1986). Effect of fertilizer source on denitrification and nitrous oxide emissions in a maize field. *Soil Science Society of America Journal* 50: 644-648.

Eichner, M.J. (1990). Nitrous oxide emissions from fertilized soils: summary of available data. *Journal of Environmental Quality* 19: 272-280.

Engels, T. and H. Kuhlmann. (1993). Effect of the rate of N fertilizer on apparent net mineralization of N during and after cultivation of cereal and sugar beet crops. *Zeitschrift für Pflanzenernährung und Bodenkunde* 156: 149-154.

Fan, M.X. and A.F. MacKenzie. (1993). Urea and phosphate interactions in fertilizer microsites: Ammonia volatilization and pH changes. *Soil Science Society of America Journal* 57: 839-845.

Fenn, L.B. and L.R. Hossner. (1985). Ammonia volatilization from ammonium or ammonium-forming nitrogen fertilizers. *Advances in Soil Science* 1: 123-169.

Ferguson, R.B., D.E. Kissel, J.K. Koelliker, and W. Basel. (1984). Ammonia volatilization from surface applied urea: Effect of hydrogen ion buffering capacity. *Soil Science Society of America Journal* 48: 578-582.

Fillery, I.R.P., P.A. Roger, and S.K. De Datta. (1986). Ammonia volatilization from nitrogen sources applied to rice fields: 2. Floodwater properties and submerged photosynthetic biomass. *Soil Science Society of America Journal* 50: 86-91.

Fillery, I.R.P., J.R. Simpson, and S.K. De Datta. (1986). Contribution of ammonia volatilization to total nitrogen loss after applications of urea to wetland rice fields. *Fertilizer Research* 8: 193-202.

Flessa, H., W. Pfau, P. Dörsch, and F. Beese. (1996). The influence of nitrate and ammonium fertilization on N_2O release and CH_4 uptake of a well-drained topsoil demonstrated by a soil microcosm experiment. *Zeitschrift für Pflanzenernährung und Bodenkunde* 159: 499-503.

Fox, R.H. and L.D. Hoffman. (1981). The effect of N fertilizer source on grain yield, N uptake, soil pH, and lime requirement in no-till corn. *Agronomy Journal* 73: 891-895.

Fox, R.H., W.P. Piekielek, and K.E. Macneal. (1996). Estimating ammonia volatilization losses from urea fertilizers using a simplified micrometeorological sampler. *Soil Science Society of America Journal* 60: 596-601.

Francis, D.D., J.W. Doran, and R.D. Lohry. (1993). Immobilization and uptake of nitrogen applied to corn as starter fertilizer. *Soil Science Society of America Journal* 57: 1023-1026.

Freney, J.R., D.L. Chen, A.R. Mosier, I.J. Rochester, G.A. Constable, and P.M. Chalk. (1993). Use of nitrification inhibitors to increase fertilizer nitrogen recovery and lint yield in irrigated cotton. *Fertilizer Research* 34: 37-44.

Freney, J.R., O.T. Denmead, I. Watanabe, and E.T. Craswell. (1981). Ammonia and nitrous oxide losses following applications of ammonium sulfate to flooded rice. *Australian Journal of Agricultural Research* 32: 37-45.

Freney, J.R., O.T. Denmead, A.W. Wood, P.G. Saffigna, L.S. Chapman, G.J. Ham, A.P. Hurney, and R.L. Stewart. (1992). Factors controlling ammonia loss from trash covered sugarcane fields fertilized with urea. *Fertilizer Research* 31: 341-349.

Freney, J.R., D.G. Keerthisinghe, S. Phongpan, P. Chaiwanakupt, and K.J. Harrington. (1995). Effect of urease, nitrification and algal inhibitors on ammonia loss and grain yield of flooded rice in Thailand. *Fertilizer Research* 40: 225-233.

Freney, J.R., A.C.F. Trevitt, S.K. De Datta, W.N. Obcemea, and J.G. Real. (1990). The interdependence of ammonia volatilization and denitrification as nitrogen loss processes in flooded rice fields in the Philippines. *Biology and Fertility of Soils* 9: 31-36.

Frye, W.W. (1977). Fall- vs. spring-applied sulfur-coated urea, uncoated urea, and sodium nitrate for corn. *Agronomy Journal* 69: 278-282.

Gentry, L.E. and F.E. Below. (1993). Maize productivity as influenced by form and availability of nitrogen. *Crop Science* 33: 491-497.

Gerendás, J., Z. Zhu, R. Bendixen, R.D. Ratcliffe and B. Sattelmacher. (1997). Physiological and biochemical processes related to ammonium toxicity in higher plants. *Zeitschrift für Pflanzenernährung und Bodenkunde* 160: 239-251.

Gooding, M.J. and W.P. Davies. (1992). Foliar urea fertilization of cereals: a review. *Fertilizer Research* 32: 209-222.

Granli, T. and O.C. Bøckman. (1994). Nitrous oxide from agriculture. *Norwegian Journal of Agricultural Sciences* Suppl. No. 12: 7-124.

Green, C.J., A.M. Blackmer, and N.C. Yang. (1994). Release of fixed ammonium during nitrification in soils. *Soil Science Society of America Journal* 58: 1411-1415.

Guiraud, G., C. Marol, and J.C. Fardeau. (1992). Balance and immobilization of $(^{15}NH_4)_2SO_4$ in a soil after the addition of Didin as a nitrification inhibitor. *Biology and Fertility of Soils* 14: 23-29.

Hamid, A. and M. Ahmad. (1993). Priming effects of ^{15}N-labelled ammonium nitrate on uptake of soil N by wheat (*Triticum aestivum* L.) under field conditions. *Biology and Fertility of Soils* 15: 297-300.

Hamid, A. and M. Ahmad. (1995). Interaction of ^{15}N-labelled ammonium nitrate, urea and ammonium sulphate with soil N during growth of wheat (*Triticum aestivum* L.) under field conditions. *Biology and Fertility of Soils* 20: 185-189.

Hargrove, W.L., D.E. Kissel, and L.B. Fenn. (1977). Field measurements of ammonia volatilization from surface applications of ammonium salts to a calcareous soil. *Agronomy Journal* 69: 473-476.

Heberer, J.A. and F.E. Below. (1989). Mixed nitrogen nutrition and the productivity of wheat in hydroponics. *Annals of Botany* 63: 643-649.

Huber, D.M., H.L. Warren, D.W. Nelson, C.Y. Tsai, and G.E. Shaner. (1980). Response of winter wheat to inhibiting nitrification of fall-applied nitrogen. *Agronomy Journal* 72: 632-637.

Hütsch, B.W., P. Russell, and K. Mengel. (1996). CH_4 oxidation in two temperate arable soils as affected by nitrate and ammonium application. *Biology and Fertility of Soils* 23: 86-92.

Izaurralde, R.C., D.E. Kissel, and M.L. Cabrera. (1990). Simulation model of banded ammonia in soils. *Soil Science Society of America Journal* 54: 917-922.

Jansson, S.L., M.J. Hallam, and W.V. Bartholomew. (1955). Preferential utilization of ammonium over nitrate by microorganisms in the decomposition of oat straw. *Plant and Soil* 4: 382-390.

Jenkinson, D.S., R.H. Fox, and J.H. Rayner. (1985). Interaction between fertilizer nitrogen and soil nitrogen–the so-called "priming" effect. *Journal of Soil Science* 36: 425-444.

Jordan, C. (1989). The effect of fertiliser type and application rate on denitrification losses from cut grassland in Northern Ireland. *Fertilizer Research* 19: 45-55.

Juma, N.G. and E.A. Paul. (1983). Effect of a nitrification inhibitor on N immobilization and release of ^{15}N from nonexchangeable ammonium and microbial biomass. *Canadian Journal of Soil Science* 63: 167-175.

Keerthisinghe, D.G., J.R. Freney, and A.R. Mosier. (1993). Effect of wax-coated calcium carbide and nitrapyrin on nitrogen loss and methane emission from dry-seeded flooded rice. *Biology and Fertility of Soils* 16: 71-75.

Keerthisinghe, D.G., X.-J. Lin, Q.-X. Luo, and A.R. Mosier. (1996). Effect of encapsulated calcium carbide and urea application methods on denitrification and N loss from flooded rice. *Fertilizer Research* 45: 31-36.

Kitur, B.K., M.S. Smith, R.L. Blevins, and W.W. Frye. (1984). Fate of ^{15}N depleted ammonium nitrate applied to no-tillage and conventional tillage corn. *Agronomy Journal* 76: 240-242.

Leyshon, A.J., C.A. Campbell, and F.G. Warder. (1980). Comparison of the effects of NO_3- and NH_4-N on growth, yield, and yield components of Manitou wheat and Conquest barley. *Canadian Journal of Plant Science* 60: 1063-1070.

Liang, B.C. and A.F. MacKenzie. (1994). Fertilization rates and clay fixed ammonium in two Quebec soils. *Plant and Soil* 163: 103-109.

Lindau, C.W., R.D. DeLaune, W.H. Patrick Jr., and P.K. Bollich. (1990). Fertilizer effects on dinitrogen, nitrous oxide, and methane emissions from lowland rice. *Soil Science Society of America Journal* 54: 1789-1794.

Magalhaes, A.M.T., P.M. Chalk, and W.M. Strong. (1984). Effect of nitrapyrin on nitrous oxide emission from fallow soils fertilized with anhydrous ammonia. *Fertilizer Research* 5: 411-422.

Malhi, S.S. (1995). Influence of source, time and method of application, and simulated rainfall on recovery of nitrogen fertilizers applied to bromegrass. *Fertilizer Research* 41: 1-10.

Malzer, G.L., K.A. Kelling, M.A. Schmitt, R.G. Hoeft, and G.W. Randall. (1989). Performance of dicyandiamide in the North Central States. *Communications in Soil Science and Plant Analysis* 20: 2001-2022.

Minami, K. (1995). The effect of nitrogen fertilizer use and other practices on methane emission from flooded rice. *Fertilizer Research* 40: 71-84.

Nommik, H. and K. Vahtras. (1982). Retention and fixation of ammonium and ammonia in soils. In *Nitrogen in Agricultural Soils*, ed. F.J. Stevenson. Madison, Wisconsin, USA: American Society of Agronomy, Crop Science Society of America, Soil Science Society of America, pp. 123-171.

Owens, L.B. (1987). Nitrate leaching losses from monolith lysimeters as influenced by nitrapyrin. *Journal of Environmental Quality* 16: 34-38.

Pan, W.L., E.J. Kamprath, R.H. Moll, and W.A. Jackson. (1984). Prolificacy in corn:

Its effects on nitrate and ammonium uptake and utilization. *Soil Science Society of America Journal* 48: 1101-1106.

Peoples, M.B., J.R. Freney, and A.R. Mosier. (1995). Minimizing gaseous losses of nitrogen. In *Nitrogen Fertilization in the Environment*, ed. P.E. Bacon. New York, USA: Marcel Dekker, pp. 565-602.

Phongpan, S., J.R. Freney, D.G. Keerthisinghe, and P. Chaiwanakupt. (1995). Use of phenylphosphorodiamidate and N-(n-butyl)thiophosphorictriamide to reduce ammonia loss and increase grain yield following application of urea to flooded rice. *Fertilizer Research* 41: 59-66.

Prakasa, E.V.S. and K. Puttanna. (1987). Nitrification and ammonia volatilisation losses from urea and dicyandiamide-treated urea in a sandy loam soil. *Plant and Soil* 97: 201-206.

Rao, A.C.S., J.L. Smith, J.F. Parr, and R.I. Papendick. (1992). Considerations in estimating nitrogen recovery efficiency by the difference and isotopic dilution methods. *Fertilizer Research* 33: 209-217.

Recous, S. and B. Mary. (1990). Microbial immobilization of ammonium and nitrate in cultivated soils. *Soil Biology and Biochemistry* 22: 913-922.

Riga, A., V. Fischer, and H.J. Van Praag. (1980). Fate of fertilizer nitrogen applied to winter wheat as $Na^{15}NO_3$ and $(^{15}NH_4)_2SO_4$ studied in microplots through a four-course rotation: 1. Influence of fertilizer splitting on soil and fertilizer nitrogen. *Soil Science* 130: 88-99.

Sahrawat, K.L. and D.R. Keeney. (1986). Nitrous oxide emissions from soils. *Advances in Soil Science* 4: 103-148.

Sanchez, C.A. and A.M. Blackmer. (1988). Recovery of anhydrous ammonia-derived nitrogen-15 during three years of corn production in Iowa. *Agronomy Journal* 80: 102-108.

Savant, N.K. and S.K. De Datta (1982). Nitrogen transformations in wetland rice soils. *Advances in Agronomy* 35: 241-302.

Scherer, H.W. and K. Mengel. (1986). Importance of soil type on the release of nonexchangeable NH_4^+ and availability of fertilizer NH_4^+ and NO_3^-. *Fertilizer Research* 8: 249-258.

Schmidt, E.L. (1982). Nitrification in soil. In *Nitrogen in Agricultural Soils*, ed. F.J. Stevenson. Madison, Wisconsin, USA: American Society of Agronomy, Crop Science Society of America, Soil Science Society of America, pp. 253-288.

Schnier, H.F. (1995). Significance of timing and method of N fertilizer application for the N-use efficiency in flooded tropical rice. *Fertilizer Research* 42: 129-138.

Schnier, H.F., M. Dingkuhn, S.K. De Datta, E.P. Marqueses, and J.E. Faronilo. (1990). Nitrogen-15 balance in transplanted and direct-seeded flooded rice as affected by different methods of urea application. *Biology and Fertility of Soils* 10: 89-96.

Sen, S. and P.M. Chalk. (1994). Solubilization of soil organic N by alkaline-hydrolysing N fertilizers. *Fertilizer Research* 38: 131-139.

Smiciklas, K.D. and F.E. Below. (1992). Role of nitrogen in determining yield of field-grown maize. *Crop Science* 32: 1220-1225.

Smith, C.J., J.R. Freney, R.R. Sherlock, and I.E. Galbally. (1991). The fate of urea nitrogen applied in a foliar spray to wheat at heading. *Fertilizer Research* 28: 129-138.

Sommer, S.G. and C. Jensen. (1994). Ammonia volatilization from urea and ammoniacal fertilizers surface applied to winter wheat and grassland. *Fertilizer Research* 37: 85-92.

Strong, W.M. (1995). Nitrogen fertilization of upland crops. In *Nitrogen Fertilization in the Environment*, ed. P.E. Bacon. New York, USA: Marcel Dekker, pp. 129-169.

Terman, G.L. (1979). Volatilization losses of nitrogen as ammonia from surface-applied fertilizers, organic amendments, and crop residues. *Advances in Agronomy* 31: 189-223.

Thomson, C.J., H. Marschner, and V. Römheld. (1993). Effect of nitrogen fertilizer form on the pH of the bulk soil and rhizosphere, and on the growth, phosphorus and micronutrient uptake by bean. *Journal of Plant Nutrition* 16: 493-506.

Timmons, D.R. (1984). Nitrate leaching as influenced by water application level and nitrification inhibitors. *Journal of Environmental Quality* 13: 305-309.

Timmons, D.R. and R.M. Cruse. (1990). Effect of fertilization method and tillage on nitrogen-15 recovery by corn. *Agronomy Journal* 82: 777-784.

Tomar, J.S. and R.J. Soper (1981). Fate of tagged urea N in the field with different methods of N and organic matter placement. *Agronomy Journal* 73: 991-995.

Vlek, P.L.G. and E.T. Craswell (1979). Effect of N source and management on ammonia volatilization losses from flooded rice-soil systems. *Soil Science Society of America Journal* 43: 352-358.

Von Rheinbaben, W. (1990). Nitrogen losses from agricultural soils through denitrification–a critical evaluation. *Zeitschrift für Pflanzenernährung und Bodenkunde* 153: 157-166.

Walters, D.T. and G.L. Malzer. (1990a). Nitrogen management and nitrification inhibitor effects on nitrogen-15 urea: 1. Yield and fertilizer use efficiency. *Soil Science Society of America Journal* 54: 115-122.

Walters, D.T. and G.L. Malzer (1990b). Nitrogen management and nitrification inhibitor effects on nitrogen-15 urea: 2. Nitrogen leaching and balance. *Soil Science Society of America Journal* 54: 122-130.

Watson, C.J., P. Poland, H. Miller, M.B.D. Allen, M.K. Garrett, and C.B. Christianson. (1994). Agronomic assessment and 15N recovery of urea amended with the urease inhibitor nBTPT (N-(n-butyl) thiophosphoric triamide) for temperate grassland. *Plant and Soil* 161: 167-177.

Whitehead, D.C. and N. Raistrick. (1990). Ammonia volatilization from five nitrogen compounds used as fertilizers following surface application to soils. *Journal of Soil Science* 41: 387-394.

Wickramasinghe, K.N., G.A. Rodgers, and D.S. Jenkinson. (1985). Transformations of nitrogen fertilizers in soil. *Soil Biology and Biochemistry* 17: 625-630.

Wiesler, F. (1997). Agronomical and physiological aspects of ammonium and nitrate nutrition of plants. *Zeitschrift für Pflanzenernährung und Bodenkunde* 160: 227-238.

Yadvinder-Singh, S.S. Malhi, M. Nyborg, and E.G. Beauchamp. (1994). Large granules, nests or bands: Methods of increasing efficiency of fall-applied urea for small cereal grains in North America. *Fertilizer Research* 38: 61-87.

Yibirin H., J.W. Johnson, and D. Eckert. (1996). Corn production as affected by daily fertilization with ammonium, nitrate and phosphorus. *Soil Science Society of America Journal* 60: 512-518.

Zhu, Z.L., G.X. Cai, J.R. Simpson, S.L. Zhang, D.L. Chen, A.V. Jackson, and J.R. Freney. (1989). Processes of nitrogen loss from fertilizers applied to flooded rice fields on a calcareous soil in north-central China. *Fertilizer Research* 18: 101-115.

Zöttl, H.W. (1990). Remarks on the effects of nitrogen deposition to forest ecosystems. *Plant and Soil* 128: 83-89.

SUBMITTED: 02/27/97
ACCEPTED: 06/10/97

The Role of Nitrogen Fixation in Crop Production

Graham W. O'Hara

SUMMARY. Biological nitrogen fixation is an important process for agricultural productivity in many cropping systems because of direct inputs of atmospheric nitrogen, and rotational effects such as disease control. Advances in molecular biology techniques provide new opportunities to understand the ecology of root nodule bacteria and may improve the selection of elite strains for inoculation. An understanding of the genetic basis of nodulation in grain and pasture legumes may improve inoculation technologies. Temperate and tropical pastures may be improved through effective inoculation, removal of nutritional constraints, and use of alternate legume species. Increases in nitrogen fixation in crop legumes may result from addressing problems in the legume host, the microsymbiont and the environment. *[Article copies available for a fee from The Haworth Document Delivery Service: 1-800-342-9678. E-mail address: getinfo@haworthpressinc.com]*

KEYWORDS. Fixation, legumes, nitrogen, nodulation, pasture, pulses, soybean

INTRODUCTION

Nitrogen is often considered to be the most important nutrient supplied from the soil for crop production (Vance, 1997; Van Kammen,

---

Graham W. O'Hara, Centre for Rhizobium Studies, School of Biological Sciences and Biotechnology, Division of Science, Murdoch University, Murdoch WA 6150, Australia (E-mail: gohara@central.murdoch.edu.au).

[Haworth co-indexing entry note]: "The Role of Nitrogen Fixation in Crop Production." O'Hara, Graham W. Co-published simultaneously in *Journal of Crop Production* (Food Products Press, an imprint of The Haworth Press, Inc.) Vol. 1, No. 2 (#2), 1998, pp. 115-138; and: *Nutrient Use in Crop Production* (ed: Zdenko Rengel) Food Products Press, an imprint of The Haworth Press, Inc., 1998, pp. 115-138. Single or multiple copies of this article are available for a fee from The Haworth Document Delivery Service [1-800-342-9678, 9:00 a.m. - 5:00 p.m. (EST). E-mail address: getinfo@haworthpressinc. com].

1997). In many agricultural situations the availability of a suitable source of nitrogen is the major factor limiting crop productivity (Bohlool et al., 1992; Peoples, Herridge, and Ladha, 1995). Inputs of nitrogen into agricultural systems are primarily from chemical fertilisers and nitrogen derived from atmospheric dinitrogen by the process of biological nitrogen fixation, which is the microbial conversion of atmospheric dinitrogen gas into plant-useable ammonia (Parsons, 1997). Indeed, the provision of biologically-fixed nitrogen plays a key role in crop production in world agriculture.

There is an increasing awareness in many areas that the development of ecologically sustainable agricultural systems is essential for maintaining agricultural productivity at sufficient levels to meet increasing demands from the growing world population. One of the key factors for sustained agricultural productivity is effective management of nitrogen in the environment (Vance, 1997). Indeed, successful manipulation of nitrogen inputs through the use of biologically-fixed nitrogen often results in farming practices that are economically viable and environmentally prudent (Bohlool et al., 1992; Giller and Cadisch, 1995; Vance, 1997). The symbiotic associations between leguminous plants and root nodule bacteria have been estimated to fix approximately 80% of the biologically-fixed nitrogen in agricultural areas (Burns and Hardy, 1975; Vance, 1997), with the remainder being contributed by a diversity of other symbiotic systems, non-symbiotic associations between nitrogen-fixing bacteria and roots, and free-living microorganisms.

The important nitrogen-fixing leguminous species have an integral role in a diverse range of cropping systems throughout agricultural areas of the world. Historically, biological nitrogen fixation has been an essential component of many farming systems for considerable periods, with evidence for the agricultural use of legumes dating back more than 4,000 years (Fred, Baldwin, and McCoy, 1932). However, farmers throughout the world continually face the problem of supplying sufficient inputs of nitrogen to maintain crop productivity and assist in the development of long-term sustainable systems (Peoples, Herridge, and Ladha, 1995). During the past fifty years the widespread use of chemical fertilisers to supply nitrogen has had a substantial impact on food production, and fertiliser nitrogen has become a major input in crop production around the world (Brown, Flavin, and Postel, 1994). However, the economic costs and frequent deleterious environmental consequences of the heavy use of chemical nitrogen fertilisers in agriculture are a global concern (Bohlool et al., 1992; Vance, 1997).

BIOLOGICAL NITROGEN FIXATION IN AGRICULTURAL SYSTEMS

During the past 30-40 years many authors have discussed the imperative need for shifting the balance of nitrogen inputs in agriculture from fertiliser-derived nitrogen to nitrogen derived from a biological source to meet the nitrogen nutritional requirements of crop plants (Evans, 1975; Peoples and Craswell, 1992; Vance, 1997). Several important objectives have been proposed by these authors for future directions of agricultural research based on the need to increase the contribution from biological nitrogen fixation to nitrogen inputs in agricultural systems. These objectives currently include the need for optimisation of the nitrogen-fixing potential of current agricultural systems, the introduction and expansion of the capacity to fix nitrogen to new farming systems and, in the longer term, the development of new crop plants with the capacity to fix their own nitrogen. In addition, Bantilan and Johansen (1995) have recently advocated increased care be exercised when researchers are developing research proposals aimed at improving well-established management options, such as *Rhizobium* inoculation technology, and exploiting the unrealised potential for genetic enhancement of the host legumes' capacity to fix nitrogen. These authors have documented the relatively poor adoption record by farmers in many countries of long-established technologies, such as inoculation using root nodule bacteria. They suggested the use of ex-ante impact analyses for development of new research proposals, with a careful estimation of benefit/cost ratios. These analyses of intended research proposals need to address topics such as the actual gains to be expected from improving and managing nitrogen from biologically-fixed sources, the extent to which the legumes can meet their nitrogen needs through fixation, the residual value of the legume-fixed nitrogen, and other benefits from using legumes in rotational, ley and phase farming systems.

Two major agronomic benefits from symbiotic biological nitrogen fixation are the direct input of atmospheric nitrogen to the agricultural system and the control of crop diseases as a consequence of involving legumes in crop rotations. The economic benefits of nitrogen fixation are generally considered to include the provision of a cheaper form of nitrogen for the crop, even taking into account the costs required for root nodule bacterial inoculants and chemical fertilisers. Other economic benefits come from the provision of a cheaper form of disease control using legume rotations rather than sole reliance on chemical fungicides and pesticides.

The environmental benefits from using biological nitrogen fixation are seen to be associated with the replacement of chemical-based technologies with a biological system. These advantages include the decreased inputs of fertiliser nitrogen resulting in decreased levels of ground-water pollution by nitrates and reduced outputs of greenhouse gas production because the process of fixing nitrogen biologically does not depend on fossil fuel. In addition, the reduced reliance on chemically-based methods for disease control will decrease the often serious environmental consequences from the use of fungicides, pesticides and herbicides.

The challenge for agricultural scientists and farmers in the future is to maximise the efficient and effective use of biological nitrogen fixation in agricultural systems. This will involve increasing the input of nitrogen from biologically-fixed sources and also effectively using this fixed nitrogen to minimise losses of nitrogen due to mineralisation of organic nitrogen from legumes and subsequent nitrification and leaching of the nitrate. Also, the losses of nitrogen from many agricultural systems due to the denitrification of nitrate need to be understood and managed.

Biological nitrogen fixation is a process only carried out by prokaryotic microorganisms, and this process has never been shown to occur in eukaryotic organisms. All of the currently known nitrogen-fixing organisms are either Eubacteria (true bacteria) or Archaeobacteria. Currently approximately 50 genera of Proteobacteria are known to fix atmospheric nitrogen (Young, 1992). Many nitrogen-fixing bacteria can achieve nitrogen fixation on their own as free-living heterotrophic and autotrophic organisms, whereas others must establish a symbiotic relationship with a eukaryote host to support nitrogen fixation. The capacity to fix nitrogen in a symbiotic association with plants is found in three major groups of microbes: the root nodule bacteria (*Rhizobium, Bradyrhizobium, Sinorhizobium, Azorhizobium*), actinomycetes (*Frankia*) and cyanobacteria (*Anabaena, Nostoc*) (Young, 1992, 1996; Young and Haukka, 1996). The major amount of fixed nitrogen is contributed by legume symbioses, with the other symbiotic systems also providing significant inputs of fixed nitrogen to agricultural systems. Nitrogen-fixing legumes are significant components of many agricultural systems, and nodulated legumes contribute the majority of the biologically-fixed nitrogen supplied to both temperate and tropical agricultural systems. Oilseed legumes occupy about 6% of land currently under cultivation and the pulse legumes are sown on about 5% of cultivated land. In addition, about 14% of the world's agricultural land is covered by temporary pastures and fodder crops.

Biological nitrogen fixation has been demonstrated to have many agronomic, economic and environmental benefits for agriculture that continue to be expounded by its advocates. These benefits include the direct supply of atmospheric nitrogen to the legume crops and pastures and the subsequent input of this nitrogen into the other components of the farming system, and the significant advantages of using legume rotations in cropping systems. This chapter provides a critical analysis of current knowledge on the role of biological nitrogen fixation in crop production. The rapidly expanding level of information on the physiological and biochemical aspects of biological nitrogen fixation is discussed in relation to ways that the new knowledge base can be applied for strategic research for improving the essential role for biological nitrogen fixation in crop production. There is an emphasis on the principal role of symbiotic legumes in temperate and tropical agricultural systems, with additional analysis of the contribution of forage legumes. The diversity of free-living bacteria and symbiotic associations contributing fixed nitrogen in rice-based systems are discussed. Finally, there is discussion of some of the promising areas for using conventional and molecular approaches to develop new crops for future exploitation of biological nitrogen fixation in crop production.

BIOCHEMICAL AND PHYSIOLOGICAL PROCESSES IN SYMBIOTIC LEGUMES

Symbiotic legumes play a key role in the maintenance of world crop production through the inputs of biologically-fixed nitrogen. Considerable strategic and applied research efforts by many agricultural scientists are currently aimed at enhancing this essential aspect of the use of legumes. In addition, the symbiosis between legumes and root nodule bacteria is being studied at a fundamental level because these associations provide excellent model systems to help develop an understanding of the molecular basis of plant-microbe interactions. As a consequence of this interest during the past decade, there has been a very rapid expansion in our knowledge of symbiotic associations at the molecular level. The development of the symbiosis between root nodule bacteria and their specific host legume is a continuous process dependent upon a highly regulated chemical communication between the two partners. There is a very high level of specificity in these symbioses. For example the fast growing rhizobia such as *Sinorhizobium meliloti* and *Rhizobium leguminosarum* have a narrow host range, with the former able to form nitrogen-fixing nodules only with *Medicago, Melilotus* and *Trigonella* species and the latter with *Pisum* and *Vicia*. The development of nitrogen-fixing nodules on legumes takes place in a continuous series of stages controlled by genes located in the bacterium as well as in the host plant. Current advances in our knowledge of the genetic basis of the legume symbiosis are providing the foundation for the development of more effective and efficient nitrogen-fixing crops and pastures.

The root nodule bacteria are common soil bacteria and they live and survive in the soil as free-living heterotrophic organisms. The study of the microbial ecology of root nodule bacteria in their natural soil environment is attracting more attention from some researchers through a growing recognition of the need to understand the role of indigenous populations of root nodule bacteria in agricultural systems involving nodulated legumes. This is an exciting area of research for microbiologists, especially with the recent developments in molecular technologies. There should be rapid advances in this area through studies using these new techniques to determine the role and effect of indigenous populations of root nodule bacteria on the success or otherwise of inoculation programs. Several studies have already demonstrated the advantages of the application of molecular approaches to overcome some of the technical problems constraining advances in our understanding of the life cycle of root nodule bacteria in the soil (Streit, Kosch, and Werner, 1992; Wilson, 1995; Wilson, Peoples, and Jefferson, 1995). Molecular microbial ecology, the application of molecular biological techniques to investigate important questions in the life and function of microbes in their environment, has been suggested to provide the opportunity to understand the ecology of root nodule bacteria in new ways that were not previously possible. The knowledge gained from these studies will hopefully assist farmers in solving the problem of delivering sufficient effective inoculant to legume roots so that they can maximise the input of biologically-fixed nitrogen into agricultural systems. Many new challenges continue to face agricultural scientists working in this area, such as the continual development and expansion of new legume species, the deterioration of symbiotic capabilities, including the loss of nodulating capacity of elite inoculant strains, and the need to develop new strains of root nodule bacteria with improved characteristics, such as greater nitrogen-fixing capacity and tolerance of environmental stresses.

There are many aspects to the whole plant level on the input of biologically-fixed nitrogen from nodulated legumes. Early effective nodulation of the legume is an important characteristic to maximise the amount of nitrogen fixed by the legume. This depends upon the successful colonisation of the legume roots by the appropriate strain of root nodule bacteria, and new data is showing that the host plant may play a key role in regulating some of the bacterial processes involved in colonisation (Phillips et al., 1997).

During the last decade molecular studies have provided us with a considerable level of understanding of the very close and intricate communication that occurs between host legume and microsymbiont. It is recognised that the efficiency of symbiotic nitrogen fixation is controlled by genes of both partners, and that the specific communication between the partners starts from the very beginning of nodule development. There has been a considerable international effort into identifying the genes of both partners which specify and regulate the formation and function of root nodules (Denarie, Debelle, and Rosenberg, 1992; Carlson, Price, and Stacey, 1994; Downie and Young, 1996; Hungria and Stacey, 1997). Not only numerous *nod* genes have been identified and characterised, but the function of many *nod* gene products is understood. Several Nod factors, the symbiotic signals produced by the bacteria, have been purified and characterised (Denarie, Debelle, and Rosenberg, 1992; Denarie, Debelle, and Prome, 1996; Long, 1996), and work is now progressing on several fronts to elucidate the signal transduction pathways leading to the regulation of the expression of the symbiotic plant genes that control the dramatic developmental switch to turn on nodule initiation inside the legume root.

Once an appropriate strain of root nodule bacteria is located in the rhizosphere of the host legume, the nodulation process begins with plant signal molecules activating rhizobia nodulation genes (*nod* genes). The expression of the *nod* genes subsequently leads to infection of the legume root and nodule development on the host plant. The specific plant signal molecules released into the rhizosphere, usually flavonoids or isoflavonoids, are excreted in root exudates and these molecules are thought to interact with regulatory *nod*D gene products in the bacteria and activate the transcription of the structural *nod* genes, whose products synthesise the Nod factors (Denarie, Debelle, and Rosenberg, 1992; Denarie, Debelle, and Prome, 1996; Long, 1996). The evidence from several studies shows that the *nod*D genes regulate nodulation in all root nodule bacteria studied to date. In some rhizobia there are multiple copies of *nod*D that can code for different NodD proteins that are activated by different plant signals. As well, some *nod*D genes have been shown to be involved in the regulation of *exo* genes as well as *nod* genes (Fisher and Long, 1992). However, in some species the control of *nod* genes also appears to involve other regulatory systems, such as the two-component sensor-regulator systems common in many bacteria. The specific *nod*D genes present in a root nodule bacterium are an important factor in determining the host-range of the bacterium (Denarie, Debelle, and Rosenberg, 1992; Denarie, Debelle, and Prome, 1996; Long, 1996).

Nodule factors (Nod factors) are the collective name given to the symbiotic signal molecules synthesised by the root nodule bacteria in response to the activation by NodD proteins (Lerouge et al., 1990). In essence, the Nod factors are the gene products controlled by the *nod* genes. The NodD protein regulates the initial stages of the process and the structural *nod*ABC genes are key elements because their inactivation results in complete loss of the ability to nodulate (Downie and Young, 1996).

The Nod factors produced by all root nodule bacteria studied to date belong to the same chemical family, they are lipochitooligosaccharides, consisting of a backbone of β-1,4-linked N-acetyl glucosamine residues. The precursor for the formation of this oligosaccharide is the same precursor as that for the lipopolysaccharide and peptidoglycan components of the outer membrane of Gram negative bacteria (Carlson, Bhat, and Reuhs, 1992). Downie and Young (1996) have outlined how a minimal Nod factor can be synthesised by the sequential activities of NodC, NodB and NodA. Structural studies of the Nod factors excreted by different rhizobial species have shown a considerable diversity in the modifications and substitutions to the Nod factors; these structural changes to the molecule seem to determine specificity and host range for nodulation. For example, the type of acyl group present on the Nod factors is a crucial determinant of host specificity (Spaink, 1995).

The biological functions of the Nod factors seem to be involved with the regulation of several important developmental responses in the root of the host plant. These characteristic changes in root development induced by Nod factors include root hair deformation, nodule morphogenesis and the concomitant induction of several nodule-specific host-plant genes (Mylona, Pawlowski, and Bisseling, 1995). Recent evidence indicates there are specific receptors for the Nod factors located on root cells, and calcium seems to be directly involved in the signalling process within the root cells following detection of the Nod factors (Ehrhardt, Wais, and Long, 1996). Downie and Young (1996) have suggested that the nodulation signalling pathway may involve a receptor-mediated, calcium-based signalling system of a type commonly found in animal cells. The intriguing question of how bacteria acquired the ability to synthesise a molecule which can be recognised by plant signalling pathways has been posed by Downie and Young (1996). Further studies on this fundamental system will be required to provide the knowledge to answer this puzzle.

The response by the host legume to the Nod factors produced by the rhizobia has been described as a general activation of the cell cycle machinery of cortical cells (Mylona, Pawlowski, and Bisseling, 1995). It would seem likely that legume root cells contain the information for the complete developmental program of nodules, and they are turned on by the Nod factor signals from the microsymbiont. As a consequence of our rapidly increasing knowledge of the molecular and biochemical basis of the development of the legume rhizobia symbiosis, there are a range of applied challenges now open for possible exploitation. An understanding of the molecular basis for the specific signalling between legumes and their associated root nodule bacteria opens the way for the development of strategies to manage and improve the efficiency of the signal exchanges leading to nodulation. Among the many important areas for these strategies to address are the effect of stressful environmental factors (acidity, nutrient deficiencies and toxicities, adverse temperatures) on the plant-microbe signalling system and the potential for genetic modification of the plant and bacteria to improve the efficiency of the signal exchanges between the two partners of the symbiosis. In addition, there is a need to assess the feasibility of exploiting this knowledge of nodulation signals for the improvement of inoculation techniques through the control of the *nod* genes in the bacteria during inoculum preparation and use. For example, can we improve rhizobial inoculants by using *nod* gene inducers?, can we manage rhizobial competition in the soil and rhizosphere through regulation of the dialogues?

It would seem likely that there are a number of other metabolite exchanges between host and bacterium that are important for legume nodulation, for example, those involving bacterial surface components, exopolysaccharides and lipopolysaccharides (Carlson, Bhat, and Reuhs, 1992; Leigh and Walker, 1994). The important metabolic exchanges that occur between the host plant cells and the microsymbiont during the post infection and nodule development stages are unknown at present. Future studies in this area will have important links related to the mineral nutrition of the developing nodule, the carbon and nitrogen exchanges, the control of the supply of oxygen to the nodule. This further work will require the application of sophisticated techniques using the recent molecular advances with gene fusions and reporter genes. Important areas for this future research to investigate include addressing how plants sense the Nod factors, and determining whether the dialogues between the two partners can be improved. The biochemical and physiological mechanisms determining the symbiotic response to adverse effects of environmental stresses, such as acidity and nutrient deficiencies are not understood at present (Dilworth and Glenn, 1991). Future studies on the effect of the edaphic environment on the exchange of molecular signals between the bacteria and legume roots will provide important information for future developments to improve the performance of inoculants in agricultural systems.

FORAGE LEGUMES

Forage legumes are an essential component of many permanent and temporary pasture systems in temperate (Peoples, Herridge, and Ladha, 1995) and tropical environments (Giller and Wilson, 1991; Humphreys, 1991; Thomas, 1995). The contribution of the legume component to the pasture system is important for the inputs of fixed nitrogen as well as for the other factors associated with the use of legumes (e.g., weed control). Productive pastures have a substantial requirement for inputs of nitrogen, and biological nitrogen fixation makes valuable inputs of nitrogen to both temporary and permanent pastures. In many agricultural systems, farmers depend on legume nitrogen fixation by forage legumes to sustain soil nitrogen fertility. Unfortunately, the technical difficulties associated with accurately measuring nitrogen fixation under field conditions have limited the number of reliable studies on the contribution of fixed nitrogen to legume-based pastures. It is only relatively recently, with the development of ^{15}N-based methodologies, that useful estimations of the contribution of biologically-fixed nitrogen in legume pastures have become possible. However, considerable progress is being made in this area recently; several studies clearly show there is little doubt that forage legumes have the potential to sustain the nitrogen requirements of temperate (Peoples Herridge, and Ladha, 1995) and tropical pastures (Thomas, 1995). The results from the relatively few reliable measurements of nitrogen fixation by pasture legumes show that levels of fixed nitrogen can be similar to those obtained from crop legumes, but in many cases only small amounts of fixed nitrogen have been measured, probably reflecting a small legume component in the pasture sward. Indeed, in many farm-based measurements in temperate and tropical pastures, there have been low values recorded for the proportion of nitrogen coming from symbiotic nitrogen fixation. Differences between the amounts of fixed nitrogen measured from experimental and farm-based investigations have been attributed to various factors associated with experimental trials. These factors include the short-term basis of many experimental studies, resulting in the presence of an unrepresentative level of legumes in experimental plots, and the absence of grazing pressure on the experimental sites (Peoples, Herridge, and Ladha, 1995).

There are several key questions still to be answered in relation to the use of nitrogen-fixing legumes in pastures. These questions cover issues including the amounts of biologically-fixed nitrogen needed to sustain pasture-based systems, quantifying the amount of nitrogen that is being fixed, and determining the nitrogen-fixing potential of the system and the fate of the fixed nitrogen in the soil, plant and animal components of the systems. The provision of data to address these issues will form the basis of an understanding of how the systems can be managed to provide sustainable legume/grass pastures.

When considering the role of nitrogen-fixing legumes in pasture and forage agriculture, an important factor is the amount of legume required in a pasture system to maximise production and live-weight gain in grazing animals, and to provide sufficient fixed nitrogen to balance the nitrogen removed or lost during cycling through the soil-plant-animal system. The amount of legume required in the pasture to provide maximum production and sufficient nitrogen will depend on the proportion of legume, plant-nitrogen derived from fixation and the total biomass of legume in the pasture. The most important characteristic determining net inputs of fixed nitrogen in a pasture system will be the maintenance of a high legume content within the pasture, and the persistence of the legume component over time (Peoples, Herridge, and Ladha, 1995). Other factors reported to affect the levels of fixation in pastures include grazing (Thomas, 1995) and the presence of localised areas of N-rich excreta from grazing animals (Ledgard and Steele, 1992).

Management strategies to maximise the legume component of pastures have been developed for pastures in cropping zones and permanently grazed areas. In essence, these strategies aim to either increase the numbers of effective rhizobia in the soil, reduce the levels of soil nitrate, decrease legume sensitivity to soil nitrate, or increase the potential for legume growth. The ultimate aim of these strategies is to manage legume-based pastures in diverse farming systems to provide a renewable source of nitrogen for agriculture. Important considerations for the success of these management options include the effective use of inoculation with appropriate root nodule bacteria, the removal of nutritional constraints resulting from deficiencies and/or toxicities in problem soils, and an awareness of the increasing range of alternate pasture legume species for exploitation of different environments. The persistence of the legume component in a pasture can be improved by maintaining adequate concentrations of all nutrients other than nitrogen, and the selection and introduction of more productive, disease-resistant cultivars or species (Peoples, Ladha, and Herridge, 1995).

TEMPERATE GRAIN LEGUMES

The principal grain legume crops grown in temperate regions include faba beans, common bean, field peas, vetch, chickpeas, lentils, lathyrus and lupin. In many agricultural systems, these grain legume crops will depend on symbiotic nitrogen fixation for the provision of a substantial proportion of their nitrogen. There can be considerable variation in the measured amounts of nitrogen fixed by temperate grain legumes; a wide range of environmental constraints can markedly influence the level of fixation and subsequent contribution of fixed nitrogen to the grain legume system. These environmental factors include the level of available moisture in the soil, the supply of essential nutrients such as P and K, and soil acidity. In addition, there are biological factors (such as competition), and management issues (such as sowing date) that can have a profound influence on the contribution of fixed nitrogen by the grain legume. A major issue also being considered in several research programs is acidification of the soil as a consequence of legume-based agriculture. This problem is a primary concern for legume-based cropping systems, with nitrate leaching and a greater cation/anion uptake ratio being identified as significant contributors to the acidification of topsoil and subsoil, respectively (C. Tang, pers. commun.). Thus, the management of soil nitrate and other nutrients is an essential consideration in strategies designed to limit the rate of soil acidification under legumes.

SOYBEAN

Soybean (*Glycine max* L. Merrill) is the most widely grown crop legume, with 55 million ha sown in 1992, producing an estimated 114 million tonnes of grain (Herridge and Danso, 1995). This legume is one of the major sources of food and animal feed because the crop can be grown in both temperate and tropical environments, and it is capable of producing high yields of grain that contains large amounts of protein and oil. The approximate composition of soybean is 40% protein, 21% oil, 34% carbohydrate and 5% ash (Keyser and Li, 1992). By weight, the protein yield of soybean is about twice that of meat; therefore, in areas with rapidly increasing populations, this crop is seen as an important component of local diets.

As a nodulating legume, soybean forms a nitrogen-fixing symbiosis with *Bradyrhizobium japonicum, B. elkanii, B. liaoningense*, and *Sinorhizobium fredii* (Young and Haukka, 1996). The nitrogen requirement of soybean can be met by both assimilation of mineral nitrogen from the soil and symbiotic nitrogen fixation. Although these mechanisms for acquiring nitrogen have independent pathways and control points, these two systems of acquiring nitrogen are interdependent and under almost all field conditions soybean plants will obtain their supply of nitrogen concurrently from both sources.

There have been many reports indicating a wide variation in the level of biological nitrogen fixation in soybean crops in field situations. The data from these studies on soybean grown in a diverse range of countries, including Australia, USA, Thailand, Canada and France, show the total amount of nitrogen fixed ranging from 0 to 450 kg ha^{-1}, with the proportion of plant nitrogen being derived from fixation varying from 0 to 95%. Some of this variation in the level of nitrogen fixation is linked to variations in crop yield. The amounts of nitrogen fixed by symbiotic soybean are affected by genetic traits of the plant (such as maturity) and environmental constraints (such as high levels of soil nitrogen), nutrient deficiencies of essential elements (such as phosphorus), adverse soil pH and water stress (Keyser and Li, 1992). However, soybean crops have the capacity to fix high levels of nitrogen, and in some studies large inputs of biologically-fixed nitrogen in farmers soybean crops have been measured. In these situations mean estimates of the proportion of plant nitrogen derived from fixation indicate a relatively high reliance on nitrogen fixation for growth and grain production. Despite this capacity of soybean for high levels of nitrogen fixation, Peoples, Herridge, and Ladha (1995) estimate that the levels of fixation achieved by soybean is less than sufficient to offset the nitrogen removed with the harvested grain. They also suggested that improving crop and soil management will have a greater impact on nitrogen fixation inputs into soybean-based systems, than genetic improvement of the host plant or symbiont.

Recent approaches to improving biological nitrogen fixation by soybean crops include attempts to optimise the numbers and effectiveness of rhizobia in the rooting zone, selection of elite strains of bradyrhizobia, improved inoculation techniques, plant breeding for promiscuous or selective nodulation, and improvement in breeding programs for nitrate tolerance in soybean. A well-documented account of the strategies for selection and breeding of soybean for enhanced nitrogen fixation is provided by Herridge and Danso (1995). They conclude that despite the considerable efforts devoted to large breeding programs in Australia, Africa and the USA, there has been an apparent lack of successful release of cultivars with improved capacity for nitrogen fixation. Two major reasons are proposed by these authors for the poor return for effort from these programs to date: firstly, the difficulty of combining a single desirable trait such as symbiotic nitrogen fixation with other important agronomic characteristics and yield traits, and secondly, the fact that accurate techniques for measuring nitrogen fixation in field-grown crops have not been widely available. While the ^{15}N-based methods are very sophisticated and can provide accurate measurements of the contribution of biological nitrogen fixation to nitrogen inputs (Herridge and Danso, 1995), the expensive equipment and time constraints associated with these assays have limited their use in many areas. The recently developed xylem ureide technique has been used in several studies (Herridge and Danso, 1995). Advances using new molecular techniques (Wilson, Peoples, and Jefferson, 1995) have provided the potential for selection and breeding of ureide-producing legumes such as soybean.

Recently, a rapid expansion of soybean production in southern Africa has been attributed to two alternate approaches for improving the level of nitrogen fixation in this grain legume crop. Large commercial farms have successfully used inoculation with locally produced *Bradyrhizobium* inoculants to increase nitrogen fixation. On the other hand, small holders have successfully used robust, promiscuously nodulating varieties of soybean developed in Zambia. These varieties nodulate with the local indigenous fast-growing rhizobia and fix nitrogen well in farmers fields. Thus, in many locations there is no requirement for inoculation to achieve adequate yields of soybean.

TROPICAL GRAIN LEGUMES

Legume pulses and oilseeds are important components of many agricultural systems in tropical regions. Biological nitrogen fixation by these legume crops has a key role in sustaining productivity in tropical agriculture. The important legume crops grown in tropical areas spread throughout the world include a diverse range of crops such as black gram, chickpea, cluster bean, common bean, cowpea, groundnut (peanut), lentil, green gram, navy beans, mung beans, and snake beans. Many studies have consistently documented benefits of crop legumes to succeeding non-legume crops (George et al., 1992; Wani, Rupela, and Lee, 1995), with improvements in yields of cereals grown after a legume crop when compared with yields obtained in cereal-cereal cropping sequences (Peoples and Craswell, 1992). The benefits from the legume crop to the system have been largely attributed by many authors to the input of biologically-fixed nitrogen (Wani, Rupela, and Lee, 1995). However, other studies have shown that in situations involving grain legumes, where the harvested grain and crop residues are removed from the field, there may be a decrease in soil nitrogen reserves (Wani, Rupela, and Lee, 1995). However, even in situations where crop legumes are depleting soil nitrogen reserves, there may still be benefits from using the crop legumes in rotations (Peoples and Craswell, 1992). Several inter-related factors may be involved in this beneficial effect of grain legumes, like local edaphic and seasonal variations, and management practices such as the rotational crop sequences. In addition, the legume crop may influence soil microbial activity, increase the availability of essential mineral nutrients, improve soil structure and other physical characteristics, and have the rotational effects on breaking disease cycles (Wani, Rupela, and Lee, 1995). As with all situations involving legumes in agricultural systems, the data from numerous studies show clearly that no general rules apply across systems; there are well documented examples where using crop legumes does not provide measurable advantages to the sustainability of the system. However, there can be no doubt that pulse and oilseed legume crops are essential components of tropical agriculture, and the challenge for scientists working in these areas is to improve biological nitrogen fixation by crop legumes, so that farmers benefit through developing sustainable cropping rotations.

The current approaches to improving biological nitrogen fixation in tropical crop legumes are focussed on one of the three principal components of symbiotic systems: the legume host, the microsymbiont and the environment (Wani, Rupela and Lee, 1995). Several research programs have been developed for the selection of tropical grain legume crops with improved symbiotic characteristics based on the documented genotypic variability for symbiotic traits, such as nodulation and nodule function (Wani, Rupela, and Lee, 1995). In addition, the approach taken with some grain legumes (breeding cultivars with greater tolerance of soil nitrate) is seen by some as a feasible approach to follow with the tropical grain legumes as well.

Important issues for improving the effectiveness of the microsymbiont component of the system all relate to the appropriate and effective use of inoculants in tropical agriculture. The first issue is the essential requirement for determining whether there is a need to inoculate in all situations where grain legumes are sown. This is still a fundamental point that has not been adequately investigated for many grain legume crops. The development of appropriate mathematical models to assist researchers, extension officers and farmers in determining the need to inoculate is a valuable exercise (Thies, Singleton, and Bohlool, 1991). There is a need for research on these models to continue, with a view to increase their accuracy for providing reliable answers in all situations (Wani, Rupela, and Lee, 1995).

Given that there is good evidence for a need to inoculate a tropical grain legume crop, then several factors will be important in influencing the success or failure of the inoculation process. The presence of established populations of root nodule bacteria in the soil may well provide strong competition for the inoculant strain in the soil and rhizosphere and at the nodulation sites on the legume root. The nature of competition in the soil and rhizosphere between naturalised and introduced strains of root nodule bacteria is poorly understood at present. Improvements in our understanding of the factors affecting competition may provide considerable benefits for improving the performance of inoculum strains of root nodule bacteria. Other factors recognised as affecting the performance of inoculum strains include the quality and stability of the inoculants, farmers having appropriate knowledge about the storage and use of inoculation technology, and the effects of adverse edaphic (acidity, alkalinity, salinity) and environmental factors (e.g., high temperature) on the inoculant strains of root nodule bacteria.

The third component being addressed by research programs relates to improvements in management practices, with a view to overcoming or alleviating adverse effects of the environment on the symbiotic grain legumes. The factors under investigation include strategies to improve the use of soil nitrogen under tropical grain legumes, and an understanding of the roles for intercropping, reduced tillage and fertilisers in enhancing the performance of tropical grain legume crops.

FLOODED RICE-BASED SYSTEMS

Current estimates show that greater than 50% of the world population is dependent on rice as the principal component of the diet. About 75% of rice is grown in wetlands, where the rice grows in flooded fields for part or all of the cropping period (Roger and Ladha, 1992). In these situations, the flooding of the soil provides for the establishment of a diverse range of nitrogen-fixing organisms that make substantial inputs of fixed nitrogen into the rice-based systems. Important nitrogen-fixing organisms present in flooded rice-based systems include heterotrophic and autotrophic free-living bacteria, photosynthetic bacteria and cyanobacteria, symbiotic cyanobacteria that associate with *Azolla* and symbiotic bacteria that form root and stem nodules on legumes. The amounts of nitrogen fixed by these diverse organisms make substantial contributions to the sustainability of wetland rice cultivation. However, estimates of the contribution to the rice-based systems of the nitrogen fixed by these organisms indicate that the majority of the nitrogen is supplied by the symbiotic associations between the aquatic fern *Azolla* and cyanobacteria from the genus *Anabaena*, and between nodule bacteria and aquatic legumes (Roger and Ladha, 1992; Watanabe and Liu, 1992; Becker, Ladha, and Ali, 1995).

The discovery of the potential as green manures of stem-nodulating, fast-growing aquatic legumes (e.g., *Sesbania rostrata*; Dreyfus and Dommergues, 1981) resulted in increased research in this novel symbiotic system. Many species of *Sesbania* and *Aeschynomene* are nodulated on their stems by specific nitrogen-fixing bacteria (Ladha et al., 1991; Watanabe and Liu, 1992) with *Sesbania rostrata* being nodulated on the stem and root by *Azorhizobium caulinodans* (Dreyfus, Garcia, and Gillis, 1988). These stem-nodulating legumes have the capacity to fix substantial amounts of nitrogen and accumulate sufficient levels of nitrogen to entirely substitute for mineral fertiliser nitrogen in the rice-based systems (Becker, Ladha, and Ottow, 1990; Becker, Ladha, and Ali, 1995).

There is a very long history in southern China and northern Vietnam of the use of *Azolla* as green manure for wetland rice-based systems (Watanabe and Liu, 1992). However, at present it is estimated that *Azolla* is only used as a green manure on < 2% of the world's rice crop (Giller and Wilson, 1991), and there has been limited adoption of legume-based green manures in tropical wetland rice farming (Becker, Ladha, and Ali, 1995). In addition, there is clear evidence of a world-wide dramatic decline in the use of green manures for wetland rice during the past 30 years (Watanabe and Liu, 1992; Becker, Ladha, and Ali, 1995). This decline in use of both the *Azolla* and legume green manures has occurred despite the high potential of both of these systems for providing to the rice-based systems substantial amounts of nitrogen, of which the major proportion comes regularly from biologically-fixed nitrogen (Becker, Ladha, and Ali, 1995).

The use of nitrogen-fixing green manures has been documented to improve rice grain yields (Singh, Khind, and Singh, 1991) through effects on physical and chemical soil parameters (Becker, Ladha, and Ottow, 1988) and reduced losses of nitrogen from the system (Buresh and De Datta, 1991). However, despite these potential benefits from green manures, a range of agronomic and economic factors have been identified as the major constraints, reducing the use of green manures in wetland rice-based systems (Becker, Ladha, and Ali, 1995). Major limitations to widespread use of green manures included the lack of sufficient seeds of the appropriate green manure legumes, the high cost of establishment, cultivation and incorporation of the green manure (Garrity and Flinn, 1988), the high cost of land and labour, and the relatively low cost of mineral nitrogen fertiliser (Ali and Narciso, 1993). Although there seems little prospect for a rapid reversal of the recent reduction in the use of green manures in flooded rice-based systems, there are still promising areas for future research. The current limited role for nitrogen-fixing green manures in rice-based systems may be increased as a result of research directed into areas such as the selection of *Azolla* and legume species with improved characteristics like stress resistance, pest resistance, and increased seed production by the leguminous species. As well, research needs to address the requirement for multipurpose green manures to overcome some of the economic constraints preventing farmer use of these potentially very beneficial and sustainable biological technologies.

FUTURE PROSPECTS OF BIOLOGICAL NITROGEN FIXATION IN CROP PRODUCTION

The rapidly accelerating technical advances in areas of molecular biology are providing technologies that will enhance research aimed at increasing the impact of biological nitrogen fixation in crop production. However, the use of sophisticated technologies needs to be approached with an awareness that the results of research must eventually feed back to farmers before there can be a substantial impact on crop productivity. Giller and Cadisch (1995) favour an ecological approach to this problem, with an emphasis on the role of the legumes in the farming system. Indeed, there is considerable merit in their recommended approach because the evidence from a number of past episodes suggests that considerable caution should be used when reviewing proposals for research developments leading to novel nitrogen-fixing crops and systems.

The prospects for improving nitrogen input into agricultural systems through exploitation of biological nitrogen fixation have been discussed at length during the past 30 years. Two approaches have been followed by researchers; improvements in the presently available nitrogen-fixing systems have been profitable in the short term, while the use of genetic and chemical manipulation to develop novel nitrogen-fixing systems is an ambitious long-term strategy that promises considerable benefits if the significant biological difficulties can be overcome.

Within the current use of legumes in agricultural systems, there appears to be considerable scope for biological nitrogen fixation to provide additional inputs of nitrogen to enhance the sustainability of agricultural systems. Particular areas where improvements in nitrogen-fixing systems may be potentially useful in the short term include the selection and use of new and alternate legumes in farming systems to overcome environmental constraints such as acidity, water stress and nutrient deficiencies. Giller and Cadisch (1995) have proposed that immediate improvements in biological nitrogen fixation in agriculture can occur through research directed at overcoming environmental constraints, introducing new legumes into farming systems and the appropriate use of inoculation in situations where effective naturalised populations of root nodule bacteria are absent. Improvements over a longer time-scale are suggested to be possible through selection and breeding of the host legume for increased nitrogen fixation and nodulation promiscuity. In situations where legumes are being sown into soils containing high populations of naturalised rhizobia and inoculation is often unsuccessful and/or unnecessary, the approach to select or engineer competitive inoculant strains is considered a longer term prospect for success. More distant prospects for success are those concerning the use of genetic manipulation to improve the legume host and establish nitrogen-fixing nodules on non-legumes, including cereals. There are many challenges facing researchers in their quest to enhance the role of biological nitrogen fixation in crop production. A considerable international effort will be needed to provide the strong support required to overcome these challenges. The increasing awareness that legumes are an essential component of many farming systems is based on improvements in our understanding of the roles for biological nitrogen fixation in developing sustainable agricultural systems. The application of sound scientific principles to develop research proposals addressing the problems limiting the benefits of nitrogen fixation to crop production, together with effective communication between researchers and farmers, will assist in improving advances in this important component of agriculture.

REFERENCES

Ali, M. and J.H. Narciso. (1993). *The Perception and Reality of GM Use in Rice: An Economic Evaluation*. Los Banos, The Philippines: IRRI.

Bantilan, M.C.S. and C. Johansen. (1995). Research evaluation and impact analysis of biological nitrogen fixation. *Plant and Soil* 174: 279-286.

Becker, M., J.K. Ladha, and J.C.G. Ottow. (1988). Stem-nodulating legumes as green manure for lowland rice. *Philippine Journal of Crop Science* 13: 121-127.

Becker, M., J.K. Ladha, and J.C.G. Ottow. (1990). Growth and nitrogen fixation of two stem nodulating legumes and their effect as green manure on lowland rice. *Soil Biology and Biochemistry* 22: 1109-1119.

Becker, M., J.K. Ladha, and M. Ali. (1995). Green manure technology: potential usage, and limitations. A case study for lowland rice. *Plant and Soil* 174: 181-194.

Bohlool, B.B., J.K. Ladha, D.P. Garrity, and T. George. (1992). Biological nitrogen fixation for sustainable agriculture: a perspective. *Plant and Soil* 141: 1-11.

Buresh, R.J. and S.K. De Datta. (1991). Nitrogen dynamics and management of rice-legume cropping systems. *Advances in Agronomy* 45: 1-59.

Burns, R.C. and R.W.F. Hardy. (1975). *Nitrogen Fixation in Bacteria and Higher Plants*. Berlin, Germany: Springer-Verlag.

Brown, L.R., C. Flavin, and S. Postel, eds. (1994). *State of the World 1994*. London, UK: Earthscan Publications.

Carlson, R.W., U.R. Bhat, and B. Reuhs. (1992). *Rhizobium* lipopolysaccharides: their structure and evidence for their importance in the nitrogen-fixing symbiotic infection of their host legumes. In *Plant Biotechnology and Development*, ed. P.M. Gresshoff. Boca Raton, Florida, USA: CRC Press, pp. 33-44.

Carlson, R.W., N.P.J. Price, and G. Stacey. (1994). The biosynthesis of rhizobial lipo-oligosaccharide nodulation signal molecules. *Molecular Plant-Microbe Interactions* 7: 684-695.

Denarie, J., F. Debelle, and J.C. Prome. (1996). *Rhizobium* lipo-chitooligosaccharide nodulation factors: Signaling molecules mediating recognition and morphogenesis. *Annual Review of Biochemistry* 65: 505-535.

Denarie, J., F. Debelle, and C. Rosenberg. (1992). Signaling and host range variation in nodulation. *Annual Review of Microbiology* 46: 497-531.

Dilworth, M.J and A.R. Glenn. (1991). *Biology and Biochemistry of Nitrogen Fixation.* Amsterdam, The Netherlands: Elsevier.

Downie, J.A. and J.P.W. Young. (1996). Evolution of specificity in legume nodulating rhizobia. In *Recent Progress in Research on Symbiotic Nitrogen Fixation*, ed. J.S. Pate. The University of Western Australia, Perth: CLIMA occasional paper No. 14.

Dreyfus, B.L. and Dommergues. (1981). Nodulation of *Acacia* species by fast and slow-growing tropical strains of *Rhizobium. Applied and Environmental Microbiology* 41: 97-99.

Dreyfus B.L., L. Garcia, and M. Gillis (1988). Characterization of *Azorhizobium caulinodans* gen. nov. sp. nov. a stem nodulating nitrogen fixing bacterium isolated from *Sesbania rostrata. International Journal of Systematic Bacteriology* 38: 89-98.

Ehrhardt, D.W., R. Wais, and S. Long. (1996). Calcium spiking in plant root hairs responding to *Rhizobium* nodulation signals. *Cell* 85: 673-681.

Evans, H.J. (1975). *Enhancing Biological Nitrogen Fixation.* Washington D.C., USA: National Science Foundation.

Fisher, R.F. and S. Long. (1992). *Rhizobium*-plant signal exchange. *Nature* 357: 655-660.

Fred, E.B., I.L. Baldwin, and E. McCoy. (1932). *Root Nodule Bacteria and Leguminous Plants.* Madison, Wisconsin, USA: University of Wisconsin.

Garrity, D.P. and J.C. Flinn. (1988). Farm-level management systems for green-manure crops in Asian rice environments. In *Sustainable Agriculture: Green Manure in Rice Farming.* Los Banos, The Philippines: IRRI, pp. 111-130.

George, T., J.K. Ladha, R.J. Buresh, and D.P. Garrity. (1992). Managing native and legume-fixed nitrogen in lowland rice-based cropping systems. *Plant and Soil* 141: 69-91.

Giller, K.E. and G. Cadisch. (1995). Future benefits from biological nitrogen fixation: An ecological approach to agriculture. *Plant and Soil* 174: 255-277.

Giller K.E. and K.J. Wilson. (1991). *Nitrogen Fixation in Tropical Cropping Systems.* Wallingford, UK: CAB International.

Herridge, D.F. and S.K.A. Danso. (1995). Enhancing crop legume N_2 fixation through selection and breeding. *Plant and Soil* 174: 51-82.

Humphreys, L.R. (1991). *Tropical Pasture Utilization.* Cambridge, UK: Cambridge University Press.

Hungria, M. and G. Stacey. (1997). Molecular signals exchanged between host plants and rhizobia: Basic aspects and potential application in agriculture. *Soil Biology and Biochemistry* 29: 819-830.

Keyser, H.H. and F. Li. (1992). Potential for increasing biological nitrogen fixation in soybean. *Plant and Soil* 141: 119-135.

Ladha, J.K., R.P. Pareek, R. So, and M. Becker. (1991). Stem nodule symbiosis and its unusual properties. In *Nitrogen Fixation: Achievements and Objectives*, eds. P.M. Gresshoff, L.E. Roth, G. Stacey, and W.E. Newton. New York, USA: Chapman and Hall, pp. 633-640.

Ledgard S.F. and K.W. Steele. (1992). Biological nitrogen fixation in mixed legume/grass pastures. *Plant and Soil* 141: 137-153.

Leigh, J.A. and G.C. Walker. (1994). Exopolysaccharides of *Rhizobium*: synthesis, regulation and symbiotic function. *Trends in Genetics* 10: 63-67.

Lerouge, P., P. Roche, C. Faucher, F. Maillet, G. Truchet, J.-C. Prome, and J. Denarie. (1990). Symbiotic host specificity of *Rhizobium meliloti* is determined by a sulphated and acylated glucosamine oligosaccharide. *Nature* 344: 781-784.

Long, S.R. (1996). *Rhizobium* symbiosis: Nod factors in perspective. *The Plant Cell* 8: 1885-1898.

Mylona, P., K. Pawlowski, and T. Bisseling. (1995). Symbiotic nitrogen fixation. *The Plant Cell* 7: 869-885.

Parsons, R. (1997). Contrasting C supply, N assimilation and N transport across a range of symbiotic plants. In *Biological Fixation of Nitrogen for Ecology and Sustainable Agriculture*, eds. A. Legocki, H. Bothe, and A. Puhler. Berlin, Germany: Springer-Verlag, pp. 211-214.

Peoples, M.B. and E.T. Craswell. (1992). Biological nitrogen fixation: investments, expectations and actual contributions to agriculture. *Plant and Soil* 141: 13-39.

Peoples, M.B., D.F. Herridge, and J.K. Ladha. (1995). Biological nitrogen fixation: an efficient source of nitrogen for sustainable agricultural production? *Plant and Soil* 174: 3-28.

Peoples, M.B., J.K. Ladha, and D.F. Herridge. (1995). Enhancing legume N_2 fixation through plant and soil management. *Plant and Soil* 174: 83-101.

Phillips, D.A., W.R. Streit, H. Volpin, and C.M. Joseph. (1997). Plant regulation of root colonization by *Rhizobium meliloti*. In *Biological Fixation of Nitrogen for Ecology and Sustainable Agriculture*, eds. A. Legocki, H. Bothe, and A. Puhler. Berlin, Germany: Springer-Verlag, pp. 133-136.

Roger, P.A. and J.K. Ladha. (1992). Biological N_2 fixation in wetland rice fields: estimation and contribution to nitrogen balance. *Plant and Soil* 141: 41-55.

Singh, Y., C.S. Khind, and B. Singh (1991). Efficient management of leguminous green manures in wetland rice. *Advances in Agronomy* 45: 135-189.

Spaink, H.P. (1995). The molecular basis of infection and nodulation by rhizobia: The ins and outs of sympathogenesis. *Annual Review of Phytopathology* 33: 345-368.

Streit, W., K. Kosch, and D. Werner. (1992). Nodulation competiveness of *Rhizobium leguminosarum* bv *phaseoli* and *Rhizobium tropici* strains measured by glucuronidase (gus) gene fusions. *Biology and Fertility of Soils* 14: 140-144.

Thies, J.E., P.W. Singleton, and B.B. Bohlool. (1991). Modeling symbiotic performance of introduced rhizobia in the field by use on indices of indigenous population size and nitrogen status of the soil. *Applied and Environmental Microbiology* 57: 29-37.

Thomas, R.J. (1995). Role of legumes in providing N for sustainable tropical pasture systems. *Plant and Soil* 174: 103-118.

Van Kammen, A. (1997). Biological nitrogen fixation for sustainable agriculture. In *Biological Fixation of Nitrogen for Ecology and Sustainable Agriculture*, eds. A. Legocki, H. Bothe, and A. Puhler. Berlin, Germany: Springer-Verlag, pp. 177-178.

Vance, C.P. (1997). Enhanced agricultural sustainability through biological nitrogen fixation. In *Biological Fixation of Nitrogen for Ecology and Sustainable Agriculture*, eds. A. Legocki, H. Bothe, and A. Puhler. Berlin, Germany: Springer-Verlag, pp. 179-186.

Wani, S.P., O.P. Rupela, and K.K. Lee. (1995). Sustainable agriculture in the semi-arid tropics through biological nitrogen fixation in grain legumes. *Plant and Soil* 174: 29-49.

Watanabe, I. and C.C. Liu. (1992). Improving nitrogen-fixing systems and integrating them into sustainable rice farming. *Plant and Soil* 141: 57-67.

Wilson, K.J. (1995). Molecular techniques for the study of rhizobial ecology in the field. *Soil Biology and Biochemistry* 27: 501-514.

Wilson, K.J., M.B. Peoples, and R.A. Jefferson. (1995). New techniques for studying competition by rhizobia and for assessing nitrogen fixation in the field. *Plant and Soil* 174: 241-253.

Young, J.P.W. (1992). Phylogenic classification of nitrogen-fixing organisms. In *Biological Nitrogen Fixation*, eds. G. Stacey, R.H. Burris, and H. Evans. New York, USA: Chapman and Hall, pp. 43-86.

Young, J.P.W. (1996). Phylogeny and taxonomy of rhizobia. *Plant and Soil* 186: 45-52.

Young, J.P.W. and K.E. Haukka. (1996). Diversity and phylogeny of rhizobia. *The New Phytologist* 133: 87-94.

SUBMITTED: 02/10/98
ACCEPTED: 02/14/98

The Chemistry
and Agronomic Effectiveness
of Phosphate Fertilizers

Mike D. A. Bolland
Robert J. Gilkes

SUMMARY. Plants take up P from soil solution, so water-soluble P fertilizers are generally more effective than poorly soluble forms. The original sources of P used for agriculture were poorly soluble materials, including manures, bones, guano and phosphate rock. In contrast, highly soluble monocalcium phosphate, monoammonium phosphate and diammonium phosphate are the major compounds present in modern, manufactured solid fertilizers containing water-soluble P. This paper describes how the water-soluble P fertilizers are made, the dissolution of P in the granules of the fertilizers, the reactions of the fertilizer solution with the soil as P moves out of the granule into the soil, the agronomic effectiveness of the fertilizer in the year of application, and in the years after application (residual value). *[Article copies available for a fee from The Haworth Document Delivery Service: 1-800-342-9678. E-mail address: getinfo@haworthpressinc.com]*

KEYWORDS. Adsorption, agronomic effectiveness, ammonium phosphate fertilizers, precipitation, superphosphate, residual value

Mike D. A. Bolland, Senior Research Officer, Agriculture Western Australia, Bunbury WA 6231, Australia. Robert J. Gilkes, Professor, Soil Science and Plant Nutrition, Faculty of Agriculture, University of Western Australia, Nedlands WA 6907, Australia.

Address correspondence to: Mike D. A. Bolland, Agriculture Western Australia, P.O. Box 1231, Bunbury WA 6231, Australia (E-mail: mbolland@agric.wa.gov.au).

[Haworth co-indexing entry note]: "The Chemistry and Agronomic Effectiveness of Phosphate Fertilizers." Bolland, Mike D. A., and Robert J. Gilkes. Co-published simultaneously in *Journal of Crop Production* (Food Products Press, an imprint of The Haworth Press, Inc.) Vol. 1, No. 2 (#2), 1998, pp. 139-163; and: *Nutrient Use in Crop Production* (ed: Zdenko Rengel) Food Products Press, an imprint of The Haworth Press, Inc., 1998, pp. 139-163. Single or multiple copies of this article are available for a fee from The Haworth Document Delivery Service [1-800-342-9678, 9:00 a.m. - 5:00 p.m. (EST). E-mail address: getinfo@haworthpressinc.com].

139

INTRODUCTION

Plant roots take up P from soil solution as orthophosphate ions. Many soils that have been developed for agriculture in Australia and elsewhere were initially P-deficient, often acutely so, because soil solution contained little or no phosphate. Consequently, farmers have been gradually adding various forms of P to the soil. The original sources of P for agriculture were animal manures, vegetable materials, guano, bones and phosphate rocks (PR). These materials contain little water-soluble P and are generally not effective fertilizers for most soils, climates and plant species. They cannot provide sufficient plant-available P for profitable agricultural production. Modern agriculture mostly relies on granulated fertilizers manufactured from phosphate rock. Most P in these fertilizers is water-soluble; they are hereafter called water-soluble P fertilizers and are the major topic of this review. The water-soluble P fertilizers are made by adding sulphuric or phosphoric acid to apatite phosphate rocks or by passing ammonia through phosphoric acid.

Various liquid P fertilizers are also used in horticulture and agriculture in developed countries, but as the reactions of liquid fertilizers with soil are similar to those for P from water-soluble P fertilizers, liquid fertilizers will not be discussed further in this review. The reader is referred to Waggaman (1969), Sample, Soper, and Racz (1980), Robinson (1980), and Young and Davis (1980) for further details on the liquid fertilizers.

The water-soluble P fertilizers are widely used in horticulture and agriculture for many reasons. They are usually very effective in overcoming P deficiency and maximising profitable production regardless of soil type, climate, and plant species. The granules are easy to handle and apply by both machinery and by hand. When other nutrient elements (N, K, Mn, Cu, Zn, and Mo) are also required for optimum production, they can be incorporated into granules of the water-soluble P fertilizers and applied in one operation. The availability to farmers of diverse water-soluble P fertilizers has made it possible to develop large areas of land for agriculture in Australia and elsewhere.

Single or ordinary superphosphate (SSP, about 9% P) was the first water-soluble P fertilizer to be manufactured on a large scale by reaction of apatite phosphate rock with sulphuric acid. Apatite phosphate rocks are also used to manufacture phosphoric acid by adding sulphuric acid to apatite phosphate rock at a different acid/rock ratio to that used for manufacturing SSP (Waggaman, 1969; Robinson, 1980). The phosphoric acid is then used to manufacture more concentrated and compound fertilizers (Waggaman, 1969; Young and Davis, 1980). Triple superphosphate (TSP) is made by reacting apatite phosphate rock with phosphoric acid, and contains about twice the P present in single superphosphate. Ammonium phosphate fertilizers are manufactured by reacting ammonia with phosphoric acid and contain about twice the P of single superphosphate, together with 11-18% N. Compound fertilizers containing mixtures of N, P and K salts may be co-granulated, or are physical mixtures of compounds. These more concentrated fertilizers are cheaper to transport per unit of nutrients, and as N, P and K are commonly required to maximise agricultural production, they are increasingly used instead of single superphosphate. The single superphosphate contains about 11% S, whereas the concentrated P fertilizers contain little S. If S is also required, blends are manufactured containing single superphosphate, or compound fertilizer granules are coated with elemental S (Young and Davis, 1980).

This paper concentrates on water-soluble P fertilizers, only briefly discussing phosphate rock fertilizers and partially acidulated fertilizers (PAPR) which may be economic P fertilizers for particular environments or economic constraints (Hammond, Chien, and Mokwunye, 1986; Sale and Blair, 1989). It discusses the P compounds in the water-soluble P fertilizers and their reaction with soils, and how these influence the agronomic effectiveness of the fertilizers. Typically only 10-30% of the P applied as water-soluble P is used by plants in the year of application, with diminishing amounts being used in subsequent years. Some P is removed in agricultural produce, by leaching in some soil types, and by soil erosion, but most is retained by adsorption by the soil, where it becomes increasingly strongly bound to soil constituents. This retained P, together with the undissolved P compounds left behind in the water-soluble P granules, and the P returned to the soil as organic matter, supply P in the years after application of fertilizer. These diverse forms of residual P are utilised by plants in future years to different extents depending on solubility, microbial process and the external P requirement of the plant species. The agronomic effectiveness of this P in the years after fertilizer application represents the residual value of the fertilizer and is the final topic discussed in this review.

PHOSPHATE ROCK
AS A DIRECT APPLICATION FERTILIZER

When phosphate rock is applied to the soil, it needs to dissolve to produce water-soluble P for plant uptake. Dissolution is achieved by reaction between the phosphate rock and the soil, and partly through the direct action of plant roots and associated mycorrhizae on phosphate rock. Hydrogen ions, present in solution through supply from the surfaces of soil constituents, and secreted from roots of plants, dissolve the phosphate rock (Khasawneh and Doll, 1978). Since most of the hydrogen ions are supplied from soil constituents, phosphate rock can only dissolve in acidic soils. To maximise dissolution, phosphate rock needs to be applied as a fine powder that is incorporated into the soil to increase soil/fertilizer contact (Alston and Chin, 1974; Khasawneh and Doll, 1978). The soil needs to remain moist for most of the growing season for the dissolution reaction to occur to a sufficient extent to provide plants with P at a rate that matches their demand (Khasawneh and Doll, 1978; Kanabo and Gilkes, 1988; Bolland and Gilkes, 1990). Hydrogen ions provided by the soil and plant roots largely determine the rate and extent of phosphate rock dissolution (Khasawneh and Doll, 1978), although the capacity of the soil to adsorb P and Ca will also affect dissolution in response to mass action considerations. It is the total amount of hydrogen ions in the soil, as measured by titration with an alkali (titratable acidity), that primarily determines the rate and extent of phosphate rock dissolution in soils (Gilkes and Bolland, 1990). The soil needs to supply enough hydrogen ions to cause extensive dissolution of the phosphate rock in order for dissolved P to match the P requirements of plants. Annual plants, which are largely used for agriculture, develop primordia early in their growth; consequently, much P is required early in plant growth to ensure that maximum yield potentials are achieved. Therefore, phosphate rock dissolution needs to be rapid early in the growing season, and the soil needs to provide enough hydrogen ions at that time to achieve the desired supply of dissolved P (Gilkes and Bolland, 1990).

Different plant species utilize phosphate rock with different efficiencies. There are many possible reasons for this. In P-deficient soils, some plant species can acidify the rhizosphere, enabling these species to more effectively utilize phosphate rock than other species (Gerdemann, 1968; Hoffland, Findenegg, and Nelemans, 1989a, b; Haynes,

1992). Some plant species have root structures that enable them to use insoluble sources of P in the soil, such as phosphate rock. For example, the proteoid roots of *Proteaceae, Banksia, Lupinus albus* and *L. cosentinii* secrete acids that dissolve poorly soluble sources of P (Gardner, Parbery, and Barber, 1981, 1982a, b; Marschner, Römheld, and Cakmak, 1987; Dinkelaker, Römheld, and Marschner, 1989; Bolland, 1995; Hinsinger and Gilkes, 1995). In another study (Kumar, Gilkes, and Bolland, 1992), maize (*Zea mays*) was effective in obtaining P from phosphate rock and single superphosphate that had been in contact with soil for 6 years, wheat (*Triticum aestivum*) was intermediate, and lettuce (*Lactuca sativa*) was the least effective. This trend was consistent with the increasing external P requirement of these species (Sommer, 1936; Nishimoto, Fox, and Parvin, 1977; Fox, 1979, 1981, 1988).

Roots of many plant species are colonized by vesicular-arbuscular mycorrhizal fungi (VAM), which increase the volume of soil from which plants can obtain P (Mosse, 1973; Gerdemann, 1975; Abbott and Robson, 1977; Pairunan, Robson, and Abbott, 1980). The presence of VAM increases the ability of plants to take up P from soil fertilized with either water-soluble P or phosphate rock, but relative to non-VAM plants, the increase was the same for water-soluble P and phosphate rock (Pairunan, Robson, and Abbott, 1980).

Francolite, a mineral in which the extent of substitution of carbonate for phosphate in the apatite structure is greatest, is the most effective phosphate rock for direct application (Chien and Hammond, 1978; Khasawneh and Doll, 1978; Hammond, Chien, and Mokwunye, 1986). It is known as reactive phosphate rock, and may have the same effectiveness as water-soluble P fertilizers for suitable soils, plants and environments (Englestad, Jugsujinda, and De Datta, 1974; Hammond, 1978; Khasawneh and Doll, 1978; Leon, Fenster, and Hammond, 1986). These situations mainly occur in humid areas where soils are moist for long periods and contain abundant hydrogen ions (i.e., high titratable acidity) (Hammond, Chien, and Mokwunye, 1986).

Partially acidulated phosphate rock fertilizers are made by adding 20-50% less acid to phosphate rock than is required to make fully acidulated superphosphate (McLean and Wheeler, 1964; McLean, Wheeler, and Watson, 1965). The partially acidulated phosphate rock fertilizers consequently contain from 20-50% less water-soluble P than superphosphate. Partially acidulated phosphate rock fertilizers have been advocated for some soils, climates and plant species where they may be equally effective and cheaper alternatives to water-soluble P fertilizers (Hammond, Chien, and Mokwunye, 1986; Sale and Blair, 1989; Bolan, White, and Hedley, 1990). In some developing countries with local deposits of reactive phosphate rock, it is sometimes more profitable to use the local phosphate rock than an expensive imported water-soluble P fertilizers, even when the agronomic effectiveness of water-soluble P fertilizers is higher (Hammond, Chien, and Mokwunye, 1986). For the majority of soils, the water-soluble P component of partially acidulated phosphate rock behaves in a similar fashion to P in water-soluble P fertilizers. The unreacted apatite in the partially acidulated phosphate rock behaves in a similar fashion to phosphate rock and would contribute to the residual value of the fertilizer (Gilkes and Lim-Nunez, 1980).

WATER-SOLUBLE P FERTILIZERS

Details of the various procedures for manufacturing water-soluble P fertilizers are provided by Waggaman (1969), Robinson (1980), and Young and Davis (1980). A summary of the procedures is provided here.

Superphosphate Fertilizers

Single or ordinary superphosphate is made by adding concentrated sulphuric acid to powdered phosphate rock containing about 14% total P. The single superphosphate fertilizer contains 9 to 10% total P, 80-90% of which is water-soluble.

Triple superphosphate is made by adding concentrated phosphoric acid to phosphate rock. Triple superphosphate contains about 20% total P, 80-90% of which is water-soluble.

The water-soluble P in both single and triple superphosphate is monocalcium phosphate (MCP, $Ca(H_2PO_4)_2 \cdot H_2O$). The insoluble P includes unreacted phosphate rock, dicalcium phosphate (DCP, $CaHPO_4$ and $CaHPO_4 \cdot 2H_2O$), and various complex Fe and Al phosphates resulting from Fe and Al impurities in the phosphate rock (Lehr et al., 1967; White, 1974; Gilkes and Lim-Nunez, 1980).

The calcium sulphate in single superphosphate is predominantly anhydrite ($CaSO_4$), with minor amounts of hemihydrite ($CaSO_4 \cdot 0.5H_2O$) and gypsum ($CaSO_4 \cdot 2H_2O$) (Gilkes, 1975).

Ammonium Phosphates

The ammonium phosphates are made by passing anhydrous NH_3 through phosphoric acid. The most commonly available ammonium phosphate fertilizers are diammonium phosphate (DAP, $(NH_4)_2HPO_4$), containing about 20% P and 18% N, and monoammonium phosphate (MAP, $NH_4H_2PO_4$), containing about 22% P and 11% N. Generally, more than 90% of the P in these fertilizers is water-soluble, with the insoluble compounds being complex NH_4^+, Al and Fe phosphates (Gilkes and Mangano, 1983).

Compound NPK Fertilizers

These fertilizers are commonly made by adding K salts, usually KCl (muriate of potash), during manufacture of superphosphate and ammonium phosphate fertilizers. Generally, the constituents are co-granulated or physically mixed in the required proportions.

REACTIONS OF WATER-SOLUBLE P FERTILIZERS IN SOIL

Present day understanding of the reactions of fertilizer P with the soil have been derived from two types of studies. The first type investigates the fate of fertilizer P when granules of the water-soluble P fertilizers, or monocalcium phosphate, are added to soil. The second type measures adsorption and desorption of dissolved P added to suspensions of soils or soil minerals.

Reaction of P from Water-Soluble P Granules or Monocalcium Phosphate with Soil

The monocalcium phosphate and ammonium phosphate (NH_4-P) compounds present in granulated water-soluble P fertilizers are soluble and hygroscopic so that even in soil with a moisture tension of about 3 bars (i.e., well below field capacity) sufficient water moves

from the soil into the granule to initiate dissolution (Lawton and Vomocil, 1954; Lehr, Brown, and Brown, 1959; Kolaian and Ohlrogge,
1959). The water is drawn into the granule by capillarity and vapour
transport, vapour transport being more important in dry soils, because
a strong osmotic potential is developed between the monocalcium
phosphate or NH_4-P in the granule and the water in the adjacent soil
(Lehr, Brown, and Brown, 1959). The osmotic potential gradient present when the fertilizer solution first enters the soil gradually disap-
pears as the fertilizer solution is diluted by reacting with the soil as it
moves further into the soil by capillarity and diffusion. There are
concentration gradients for all fertilizer constituents going from the
granule into the soil, resulting in diffusion of ions from the granule
into the soil. The dissolved P reacts with dissolved Ca, Al and Fe in
soil solution, as well as Ca, Al and Fe exposed at the surfaces of soil
constituents and organic matter, thereby maintaining the concentration
gradient between the granule and the soil. The first precipitated P
compounds and adsorption complexes with soil constituents may be
metastable, so that dissolution and desorption reactions continue producing more stable, less soluble P forms. These reactions continue,
even in quite dry soils, albeit more slowly (Bell and Black, 1977a, b;
Barrow, 1983). Therefore, in time, as a consequence of the continuing
reactions, most P becomes incorporated into more stable, less soluble
forms that are less available to plants than the forms that initially
developed adjacent to granules (Barrow, 1980, 1983).

Some reactions of the dissolved P can occur in the granule before
the concentrated fertilizer solution enters the soil. Dicalcium phosphate, as DCP ($CaHPO_4$) and DCP dihydrate ($CaHPO_4 \cdot 2H_2O$), may
precipitate in superphosphate granules (Lehr, Brown, and Brown,
1959; Bouldin, Lehr, and Sample, 1960). The Fe and Al impurities in
the phosphate rock used to make the water-soluble P fertilizers may
also precipitate in the granule, forming complex Fe and Al phosphates, either during manufacture and storage of the fertilizer, or during dissolution in the soil (Lehr et al., 1967; White, 1974; Gilkes and
Lim-Nunez, 1980; Gilkes and Mangano, 1983).

Once the concentrated fertilizer solution has entered the soil, it
reacts with the soil near the granule. In the first few mm of soil, the
concentrated fertilizer solution can dissolve soil constituents, increasing concentrations in the soil solution of Si, Fe, Al, Mn, Ca, Mg and K
(Low and Black, 1948; Kitterick and Jackson, 1955, 1956; Lindsay

and Stephenson, 1959a, b). These dissolved ions participate in reactions, resulting in the precipitation of P from solution as the solubility products for various P compounds are exceeded. The nature of the compounds depends greatly on both fertilizer composition and soil properties, with a large number of compounds being reported in the literature, particularly for neutral to alkaline soils (e.g., Bell and Black, 1977a, b). These precipitated compounds continue to react with soil solution with a change to more stable, less-soluble P compounds with time and as the composition of soil solution changes (Lehr and Brown, 1958; Strong and Racz, 1970; Bell and Black, 1977a, b). Adsorption of P by the surfaces of soil constituents also takes place close to the granules at soil solution P concentrations that are near those associated with the maximum P sorption capacity of the soil constituents (i.e., low energy P sorption sites are occupied by P) (Sample, Soper, and Racz, 1980). It is often not possible to differentiate between precipitation and adsorption reactions in this zone (Sample, Soper, and Racz, 1980; Benbi and Gilkes, 1987). Precipitation may be the dominant process removing P from solution within and near the fertilizer granule because concentrations of P, Ca and other ions in solution are likely to be in excess of those for the solubility products for many P compounds (Sample, Soper, and Racz, 1980). In high-P-retaining soils containing poorly soluble minerals (e.g., many tropical soils) adsorption may dominate over precipitation. As a result of precipitation and adsorption of P from solution occurring near the granule in low-P-retaining soils containing quite soluble minerals, the concentration of P in solution rapidly falls with increasing distance from the granule, and eventually the typically low P concentrations in soil solution (< 0.2 μg ml^{-1}, Lewis and Quirk, 1965) are encountered. At these low concentrations, P retention by the soil is more likely to be due to adsorption than precipitation, because the solubility products of many P compounds are less likely to be exceeded (Sample, Soper, and Racz, 1980). The distance from the granule at which adsorption of P dominates over precipitation depends on soil and fertilizer properties, but is commonly about 10 mm from the granule. The distribution of P in soil adjacent to a fertilizer granule has been measured by Benbi and Gilkes (1987). For a high-P-retaining lateritic soil about 55% of P from a triple superphophate granule was present within 5 mm of the granule after 4 weeks in the soil. The corresponding value was 28% for a low-P-retaining sandy podzol.

Phosphate Adsorption by Soils

Much P (phosphate ion) is adsorbed by reacting with Fe, Al, Ca and other ions coordinated with oxygen and hydroxide ions exposed at the surface of soil constituents. In acid soils, the soil constituents that adsorb P include crystalline Fe and Al oxides and oxyhydroxides, clay minerals, amorphous compounds of Fe and Al that may exist as coatings on soil constituents, and Al that is associated with organic matter (Sample, Soper, and Racz, 1980; Hughes and Gilkes, 1994). Alkaline soils commonly contain carbonates which adsorb P; P is also commonly precipitated as Ca-phosphate from the alkaline, Ca-rich soil solution. Those constituents that adsorb P in acid soils also adsorb P in alkaline soils (Boischot, Coppenet, and Hebert, 1950; Cole, Olsen, and Scott, 1953; Norrish and Rosser, 1983).

Many surface chemical reactions and mechanisms have been proposed for the adsorption of P by surfaces of soil minerals, with much attention being directed to synthetic Fe and Al oxides (Hingston et al., 1968; Hingston, Posner, and Quirk, 1972; Bowden et al., 1973; Rajan, Perrott, and Saunders, 1974), and clay minerals (Muljadi, Posner, and Quirk, 1966a, b, c). Mechanisms have been proposed for adsorption by metal-organic complexes and carbonates (Holford, Wedderburn, and Mattingly, 1974; Holford and Mattingly, 1975). There is relatively little known of the actual P-sorption mechanisms of whole soils, although various processes have been postulated (Barrow, 1983, 1985, 1987). However, it is generally agreed that phosphate ions are adsorbed onto the surfaces of soil constituents when the oxygen atoms of the phosphate ion donate lone pairs of electrons to fill up the outer electron shell of metal atoms, principally Fe and Al, which are coordinated with hydroxide and oxygen ions exposed at the surfaces of soil constituents (Wild, 1950; Muljadi, Posner, and Quirk, 1966a, b, c; Hingston et al., 1968; Rajan, Perrott, and Saunders, 1974; Bowden, Posner, and Quirk, 1980). The phosphate ion replaces surface ions, including hydroxide, sulphate, molybdate and citrate ions, as well as water. This exchange occurs because the phosphate ion is more strongly adsorbed than the ion it replaces, so more stable surface compounds are formed by the adsorbed P (i.e., specific adsorption) (Hingston et al., 1968; Bowden, Posner, and Quirk, 1980; Barrow, 1985, 1987).

The adsorption of P by the surfaces of soil constituents is influenced by many factors that affect the charge on the surface of the soil constituents and the form of phosphate in solution. These factors influence the affinity of the surface for phosphate. For example, changes in the pH of soil solution result in adsorption or desorption of hydrogen ions at the soil surface. When hydrogen ions are adsorbed onto soil surfaces, they increase the positive charge of the surface (Bowden, Posner, and Quirk, 1977; Barrow et al., 1980, 1981). At the same time, the phosphate ion in soil solution becomes associated with a hydrogen ion (i.e., $HPO_4^{2-} \Rightarrow H_2PO_4^{-}$), lowering the negative charge of the ion. Therefore, adsorption of phosphate ions onto the surfaces of soils strengthens the electrostatic attraction between the positive surfaces and the negative phosphate ion, but this may be reduced by the reduced charge of the phosphate ions in solution as the ion becomes associated with hydrogen ions. Furthermore, adsorption of positive metal ion (e.g., Al^{3+}, Zn^{2+} or $ZnCl^{+}$ from soil solution) onto the surface of minerals may increase the net positive charge on the surface, which will increase the adsorption of phosphate (Bolland, Posner, and Quirk, 1977; Barrow et al., 1981). Adsorption of phosphate and other anions, such as citrate, onto mineral surfaces increases the net negative charge of the surface, reducing phosphate sorption and also results in subsequent adsorption occurring at sites of lower energy. Adsorption of cations and anions by the surfaces of synthetic oxides and soils are discussed in greater detail by Bowden, Posner, and Quirk (1980) and Barrow (1985, 1987).

AGRONOMIC EFFECTIVENESS
OF WATER-SOLUBLE P FERTILIZERS

A consequence of the complex adsorption and precipitation reactions discussed is that the agronomic effectiveness of the water-soluble P fertilizers should be greatest immediately after they dissolve and the resultant solution has entered the soil. This is when much P is present as dissolved P, as water-soluble P compounds, or as P adsorbed onto low energy sites; this P is consequently most available for plant uptake. The continuing adsorption and precipitation reactions in the soil produce progressively more stable, but less soluble P compounds. Therefore, the fertilizer P becomes less effective in supporting plant production (i.e., effectiveness decreases). As annual agricultural plants need abundant P very early in their growth so as to maximise their growth potential, it is important that fertilizer dissolution occurs at or shortly before germination.

The situation described in the previous paragraph is particularly important for soils with a low P status and high P-sorption capacity where P deficiency is a major constraint to plant production, and P fertilizers are not highly effective. As will be discussed in the next section, the continuous annual application of water-soluble P fertilizers eventually improves the P status of the soil, so that much of the P required for plant growth is derived from fertilizer P applied in previous years.

Results of field experiments mostly do not indicate a systematic difference in the agronomic effectiveness of P from superphosphate and ammonium phosphate fertilizers. It is therefore concluded that all water-soluble P fertilizers are equally effective per unit of P. Thus the agronomic effectiveness of the water-soluble P fertilizers is most strongly influenced by the capacities of the soil to retain and release P (Ozanne and Shaw, 1967). In general, the greater the capacity of the soil to retain P, the flatter the relationship between yield and the level of P applied (i.e., the yield response curve, Figure 1). Therefore, as the P retention capacity of the soil increases, a larger amount of fertilizer P is needed to produce the desired yield.

Factors other than P retention by the soil also affect the yield response curves for water-soluble P fertilizers. For example, yield response curves are likely to become flatter, so more P needs to be applied to produce the same relative yield, as the soil becomes acidic (Mengel and Kirkby, 1987). Plant roots take up less P from acidic soils (Adams and Pearson, 1967). In acid soils, increasing amounts of Al^{3+} ions that exist in soil solution reduce root growth, thereby decreasing the ability of plant roots to explore the soil and take up P (Helyar, 1978). Similarly, when plants are under stress, due to disease (Brennan, 1988, 1989), pests (Brennan and Grimm, 1992), waterlogging (Bolland and Baker, 1987) and other constraints, their capacity to utilise soil P is reduced.

Methods of applying fertilizers can greatly affect their agronomic effectiveness. Banding P fertilizers to the side and below the seed avoids P toxicity that would reduce germination of seedlings. Banding increases the agronomic effectiveness of fertilizers as developing roots are in intimate contact with the P-enriched soil adjacent to the fertilizer granule (Rudd and Barrow, 1973; Engelstad and Terman, 1980;

FIGURE 1. Calculation of residual value (RV), from the relationship between yield for superphosphate applied one year previously (SP1) or four years previously (SP4) relative to freshly applied current superphosphate (SPF), by comparing the amount of P required to produce the same yield. The RV of SP1 is 5/10 (= 0.5), and of SP4 is 5/20 (= 0.25). Therefore, SP1 is half as effective as SPF, or twice as much SP1 is required to produce the same yield as SPF.

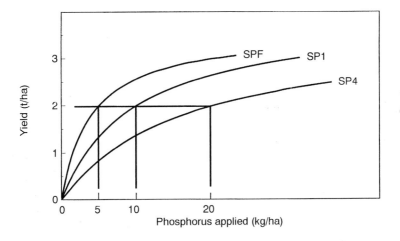

Benbi and Gilkes, 1987; Jarvis and Bolland, 1991). Concentrating the fertilizer in bands in the soil reduces the proportion of the fertilizer P that is strongly retained by the soil, so a greater proportion is available to be taken up by plant roots. Furthermore, when fertilizer is applied to the soil surface, most of the P is retained at the surface and therefore may be inaccessible to plant roots. When surface soil dries between rains during the growing season, roots are unable to utilise soil P and the agronomic effectiveness of the fertilizer is consequently decreased. Fertilizer banded at depth in the soil is more likely to be in moist soil for longer periods, increasing absorption of P by plant roots. Mixing fertilizer through the soil increases the volume of soil that reacts with dissolved P so that a greater proportion of P will be adsorbed at high energy sites, thereby reducing the proportion of the P taken up by plant roots. This effect becomes more pronounced as the capacity of the soil to retain P increases. For some crops, soils and environments, the highest yields are obtained by a combination of banded and surface-applied P (Barber, 1958; Welch et al., 1966; Ham et al., 1973). Field experiments have consistently shown that the agronomic effectiveness of water-soluble P fertilizers is greatest when the fertilizer is banded with, below or to the side of the seed while sowing crops. For pastures, water-soluble P fertilizers should be applied at or near the start of the growing season (Engelstad and Terman, 1980).

Annual plants need abundant P very early in growth. The initial source of P is the seed. The concentration of P in the seed of annual plants can affect early growth (Austin, 1966; Bolland and Baker, 1988), and sometimes final yields (Bolland and Baker, 1989; Bolland, Paynter, and Baker, 1989), irrespective of the availability of P in the soil. Therefore, farmers should use seed harvested from plants grown in soils with a high P status. For most crops, yield responses to increasing P concentration in the seed cease when the P concentration in the seed is in excess of 3 g kg^{-1} DW (0.3% P) (Bolland et al., 1990).

RESIDUAL VALUE OF FERTILIZERS

Only 10-30% of P applied as water-soluble P fertilizers is usually utilised by plants in the year of application. Most P from the fertilizer granule is retained by the soil as adsorbed-P reaction products or residual fertilizer compounds. Some of the P that is taken up by crops or pasture is returned to the soil as organic matter. In addition, the P taken up by soil animals and microorganisms is also returned to the soil as organic matter. The P compounds not taken up by plants in the year of application, and the P returned to the soil as organic matter, can be utilised in future years. That is, P fertilizers have a residual value that is of economic significance to farmers because it reduces the need for fertilizer in subsequent years. The various forms of residual P provide P to soil solution that is utilized by plants for many years after application of water-soluble P (Barrow, 1980; Bolland, Gilkes, and D'Antuono, 1988; Bolland and Gilkes, 1990). Some of P applied in fertilizers will be removed from the potentially available pool in agricultural produce, by leaching to below rooting depth, by soil erosion and/or will be retained by the soil in almost insoluble P compounds (e.g., crandallite group minerals, Norrish and Rosser, 1983).

The effectiveness of the sum of residual P forms for plant production can be determined using either plant yield or chemical soil tests for available P. As calculated by yield, the effectiveness of the residual P should be measured relative to the effectiveness of P applied in the current growing season, hereafter called current P, by comparing the amounts of residual and current P required to produce the same yield (Barrow and Campbell, 1972; Barrow, 1980). This is done by adequately defining the relationship between yield and the level of P applied for both residual and current P, by using several levels of P applied in the previous year(s) and in the current year, and by ensuring that no other nutrient deficiency, toxicity or other condition (e.g., soil pH, waterlogging, disease, pests, soil structure) limit plant yield. The residual value is then calculated by dividing the amount of current P required to produce the target yield by the amount of residual P required to produce the same yield (Figure 1).

For most soils and plant species the residual value of water-soluble P fertilizers decreases markedly in the first and second years after application (Figure 2, see also Barrow, 1980; Bolland, Gilkes, and D'Antuono, 1988). This is due to the rapid conversion of P in soil solution to more stable, less soluble compounds. Consequently, in the first year after application, the current P rapidly becomes less effective in supporting plant growth. For a single site, the residual value may vary markedly between different years (Bolland, Gilkes, and D'Antuono, 1988) due to the influence of different seasonal conditions experienced in different years on the capacity of plant roots to exploit different forms of P in the soil. That is, different seasonal conditions experienced in the different years affect the amount of plant-available P released from the various sources of residual P present in the soil, and the ability of plants growing in the soil to utilize this P to produce yield. For example, the duration of dry periods when plants are unable to exploit P in the topsoil differs between years. It is of interest to note that the residual value of poorly-soluble P fertilizers, such as phosphate rock, does not experience a similar decline in effectiveness (Figure 2, see also Barrow, 1980). However, the effectiveness of phosphate rock is commonly much less than for water-soluble P for periods up to at least 5 years after application (Figure 2).

Soil test P (available P) has been found to indicate the same trends in fertilizer residual effectiveness for water-soluble P fertilizers that are expressed by plant yield. For example, as shown in Figure 3, sodium bicarbonate soluble soil test P values increase with increasing level of P applied as single superphosphate, and decrease with increasing time from application of the fertilizer to the soil. Soil testing for P is used to estimate the amount of available P present in the soil originated from the many diverse P forms discussed above, and includes

FIGURE 2. The residual value (RV) of superphosphate and rock phosphates measured in Australian field experiments. The RV was calculated by dividing the amount of P as freshly applied (current) superphosphate required to produce a given yield by the amount of previously applied P fertilizer required to produce the same yield (see Figure 1). Data are mean RV for all published values. Vertical bars represent ± SE when greater than the symbol. For the individual experiments data are from: *Superphosphate*, Trumble and Donald (1938) □; Arndt and McIntyre (1963) ●; Barrow and Carter (1978) ○; Bolland et al. (1984) △; *apatite rock phosphate*: Arndt and McIntyre (1963) ●; Bolland and Bowden (1984) ◆ (Duchess rock phosphate/QRP1/) and ◇ (Duchess rock phosphate/QRP2/); *Calciphos* (calcium iron aluminum rock phosphates from Christmas Island heated at about 500°C): Bolland et al. (1984) △; Bolland and Bowden (1984) ◆ (C500-1) and ◇ (C500-2). Adapted from Bolland, Gilkes, and D'Antuono (1988).

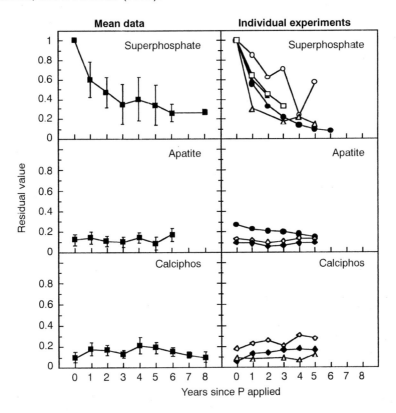

FIGURE 3. Relationship between bicarbonates soil test P (Colwell, 1963) and the amount of P applied as single superphosphate. The fertilizer was applied to the soil surface and incorporated into the top 10 cm of the soil. Soil samples to measure P were collected just after application and incorporation of fertilizer, and 156 and 869 days after application. Data of Bolland et al. (1987).

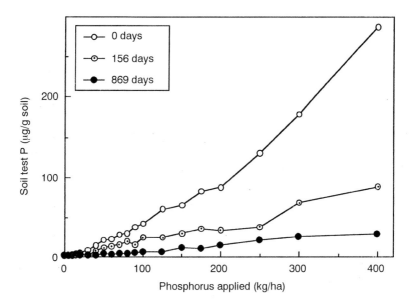

contributions from native P and fertilizer P applied in previous years (Thomas and Peaslee, 1973; Fixen and Grove, 1990). Soil test P values are related to plant yields produced later on in the next growing season through field experiments, which provide soil P test calibrations for various soil types, environments and plant species. The calibration curve indicates the yields that are likely to be produced from the current P status of the soil (i.e., from the P that is already present in the soil) and it is then possible to determine the fertilizer required to obtain a target yield. Long-term field experiments with a single plant species have been used to calibrate soil test P values at a single site and have found that soil P test calibrations are different in different years (Bolland, Gilkes, and Allen, 1988; Bolland, Allen, and Gilkes, 1989). This variation may be largely attributed to the effect of different seasonal conditions experienced in different years on the supply of P for plant uptake from the sources of residual P present in the soil and on yield response as there was no similar variation in soil test values between years, but instead a systematic decline in available P as is discussed above (Bolland, Gilkes, and Allen, 1988; Bolland, Allen, and Gilkes, 1989).

CONCLUSIONS

This brief review has shown that the availability of fertilizer P to plants can be adequately explained on the basis of the chemical reactions of soils and fertilizers, together with the interaction of seasonal conditions on the capacity of plant roots to exploit soil P.

REFERENCES

Abbott, L.K. and A.D. Robson. (1977). Growth stimulation of subterranean clover with vesicular arbuscular mycorrhizas. *Australian Journal of Agricultural Research* 28: 639-649.

Adams, F. and R.W. Pearson. (1967). Crop response to lime in southern United States and Puerto Rico. In *Soil Acidity and Liming*, eds. R.W. Pearson and F. Adams. Madison, Wisconsin, USA: American Society of Agronomy, pp. 161-206.

Alston, A.M. and K.W. Chin. (1974). Response of subterranean clover to rock phosphates as affected by particle size and depth of mixing in the soil. *Australian Journal of Experimental Agriculture and Animal Husbandry* 14: 649-655.

Arndt, W. and G.A. McIntyre. (1963). The initial and residual effects of superphosphate and rock phosphate for sorghum on lathyritic red earth. *Australian Journal of Agricultural Research* 14: 785-795.

Austin, R.B. (1966). The growth of watercress (*Rorippa nasturtium aquaticum* (L.) Hayek) from seed as affected by the phosphorus nutrition of the parent plant. *Plant and Soil* 14: 113-120.

Barber, S.A. (1958). Relation of fertilizer placement to nutrient uptake and crop yield. 1. Interaction of row phosphorus and the soil level of phosphorus. *Agronomy Journal* 50: 535-539.

Barrow, N.J. (1980). Evaluation and utilization of residual phosphorus in soils. In *The Role of Phosphorus in Agriculture*, eds. F.E. Khasawneh, E.C. Sample, and E.J. Kamprath. Madison, Wisconsin, USA: American Society of Agronomy, pp. 339-359.

Barrow, N.J. (1983). A mechanistic model for describing the sorption and desorption of phosphate by soil. *Journal of Soil Science* 34: 733-750.

Barrow, N.J. (1985). Reactions of anions and cations with variable charge soils. *Advances in Agronomy* 38: 183-230.

Barrow, N.J. (1987). *Reactions with Variable Charge Soils*. Dordrecht, The Netherlands: Martinus Nijhoff.

Barrow, N.J. and N.A. Campbell. (1972). Methods of measuring the residual value of fertilizers. *Australian Journal of Experimental Agriculture and Animal Husbandry* 12: 502-510.

Barrow, N.J. and E.D. Carter. (1978). A modified model for evaluating residual phosphate in soil. *Australian Journal of Agricultural Research* 29: 241-254.

Barrow, N.J., J.W. Bowden, A.M. Posner, and J.P. Quirk. (1980). An objective method for fitting models of ion adsorption on variable charge surfaces. *Australian Journal of Soil Research* 18: 37-47.

Barrow, N.J., J.W. Bowden, A.M. Posner, and J.P. Quirk. (1981). Describing the adsorption of copper, zinc and lead on a variable charge mineral surface. *Australian Journal of Soil Research* 19: 309-321.

Bell, L.C. and C.A. Black. (1977a). Transformation of dibasic calcium phosphate dihydrate and octacalcium phosphate in slightly acid and alkaline soils. *Soil Science Society of America Proceedings* 34: 583-587.

Bell, L.C. and C.A. Black. (1977b). Crystalline phosphates produced by interaction of orthophosphate fertilizers with slightly acid and alkaline soils. *Soil Science Society of America Proceedings* 34: 735-740.

Benbi, D.K. and R.J. Gilkes. (1987). The movement into soil of P from superphosphate grains and its availability to plants. *Fertilizer Research* 12: 21-36.

Bioschot, P., M. Coppenet, and J. Hebert. (1950). The fixation of phosphoric acid on calcium carbonate in soils. *Plant and Soil* 2: 311-322.

Bolan, N.S., R.E. White, and M.J. Hedley. (1990). A review of the use of phosphate rocks as fertilizers for direct application in Australia and New Zealand. *Australian Journal of Experimental Agriculture* 30: 297-313.

Bolland, M.D.A. (1995). *Lupinus cosentinii* more effectively utilizes low levels of phosphorus from superphosphate than *Lupinus angustifolius*. *Journal of Plant Nutrition* 18: 421-435.

Bolland, M.D.A. and M.J. Baker. (1987). Increases in soil water content decrease the residual value of superphosphate. *Australian Journal of Experimental Agriculture* 27: 571-578.

Bolland, M.D.A. and M.J. Baker. (1988). High phosphorus concentrations in seed of wheat and annual medic are related to higher rates of dry matter production of seedlings and plants. *Australian Journal of Experimental Agriculture* 28: 765-770.

Bolland, M.D.A. and M.J. Baker. (1989). High phosphorus concentration in *Trifolium balansae* and *Medicago polymorpha* seed increases herbage and seed yields in the field. *Australian Journal of Experimental Agriculture* 29: 791-795.

Bolland, M.D.A. and J.W. Bowden. (1984). The initial and residual value for subterranean clover of phosphorus from crandallite rock phosphates, apatite rock phosphates and superphosphate. *Fertilizer Research* 5: 295-307.

Bolland, M.D.A. and R.J. Gilkes. (1990). Rock phosphates are not effective fertilizers in Western Australian soils: a review of one hundred years of research. *Fertilizer Research* 22: 79-95.

Bolland, M.D.A., D.G. Allen, and R.J. Gilkes. (1989). The influence of seasonal conditions, plant species and fertilizer type on the prediction of plant yield using the Colwell bicarbonate soil test for phosphate. *Fertilizer Research* 19: 143-158.

Bolland, M.D.A., R.J. Gilkes, and D.G. Allen. (1988). The residual value of superphosphate and rock phosphates for lateritic soils and its evaluation using three soil phosphate tests. *Fertilizer Research* 15: 253-280.

Bolland, M.D.A., R.J. Gilkes, and M.F. D'Antuono. (1988). The effectiveness of rock phosphate fertilizers in Australian agriculture: a review. *Australian Journal of Experimental Agriculture* 28: 655-668.

Bolland, M.D.A., B.H. Paynter, and M.J. Baker. (1989). Increasing phosphorus concentration in lupin seed increases grain yields on phosphorus deficient soil. *Australian Journal of Experimental Agriculture* 29: 797-801.

Bolland, M.D.A., A.M. Posner, and J.P. Quirk. (1977). Zinc adsorption by goethite in the absence and presence of phosphate. *Australian Journal of Soil Research* 15: 279-286.

Bolland, M.D.A., J.W. Bowden, M.F. D'Antuono, and R.J. Gilkes. (1984). The current and residual value of superphosphate, Christmas Island C-ore, and Calciphos as fertilisers for a subterranean clover pasture. *Fertilizer Research* 5: 335-354.

Bolland, M.D.A., R.J. Gilkes, D.G. Allen, and M.F. D'Antuono. (1987). Residual value of superphosphate and Queensland rock phosphate for serradella and clover on very sandy soils as assessed by plant growth and bicarbonate-soluble phosphorus. *Australian Journal of Experimental Agriculture* 27: 275-282.

Bolland, M.D.A., M.M. Riley, B.D. Thomson, B.H. Paynter, and M.J. Baker. (1990). Seed phosphorus–its effect on plant production. *Journal of Agriculture Western Australia* (4th Series) 31: 20-22.

Bouldin, D.R., J.R. Lehr, and E.C. Sample. (1960). The effect of associated salts on transformations of monocalcium phosphate monohydrate at the site of application. *Soil Science Society of America Proceedings* 24: 464-468.

Bowden, J.W., A.M. Posner, and J.P. Quirk. (1977). Ionic adsorption on variable charge mineral surfaces. Theoretical charge development and titration curves. *Australian Journal of Soil Research* 15: 121-136.

Bowden, J.W., A.M. Posner, and J.P. Quirk. (1980). Adsorption and charging phenomena in variable charge soils. In *Soils with Variable Charge*, ed. B.K.G. Theng. Lower Hutt, New Zealand: New Zealand Society of Soil Science, pp. 146-166.

Bowden, J.W., M.D.A. Bolland, A.M. Posner, and J.P. Quirk (1973). Generalized model for anion and cation adsorption at oxide surfaces. *Nature Physical Sciences* 24: 81-83.

Brennan, R.F. (1988). Effect of phosphorus deficiency in wheat on the infection of roots by *Gaeumannomyces graminis* var. *tritici*. *Australian Journal of Agricultural Research* 39: 541-546.

Brennan, R.F. (1989). Effect of superphosphate and superphosphate plus flutriafol on yield and take-all of wheat. *Australian Journal of Experimental Agriculture* 29: 247-252.

Brennan, R.F. and M. Grimm. (1992). Effect of aphids and mites on herbage and seed production of subterranean clover (cv. Daliak) in response to superphosphate and potash. *Australian Journal of Experimental Agriculture* 32: 39-47.

Chien, S.H. and L.L. Hammond. (1978). A comparison of various laboratory methods for predicting the agronomic potential of phosphate rocks for direct application. *Soil Science Society of America Journal* 42: 935-939.

Cole, C.V., S.R. Olsen, and C.O. Scott. (1953). The nature of phosphate sorption by calcium carbonate. *Soil Science Society of America Proceedings* 17: 352-356.

Colwell, J.D. (1963). The estimation of the phosphorus fertilizer requirements of wheat in southern New South Wales by soil analysis. *Australian Journal of Experimental Agriculture and Animal Husbandry* 3: 190-197.

Dinkelaker, B., V. Römheld, and H. Marschner. (1989). Citric acid excretion and precipitation of calcium in the rhizosphere of white lupin (*Lupinus albus* L.). *Plant, Cell, and Environment* 12: 285-292.

Engelstad, O.P., A. Jugsujinda, and S.K. De Datta. (1974). Response of flooded rice to phosphate rocks varying in citrate solubility. *Soil Science Society of America Proceedings* 38: 524-529.

Engelstad, O.P. and G.L. Terman. (1980). Agronomic effectiveness of phosphate fertilizers. In *The Role of Phosphorus in Agriculture*, eds. F.E. Khasawneh, E.C. Sample, and E.J. Kamprath. Madison, Wisconsin, USA: American Society of Agronomy, pp. 311-332.

Fixen, P.E. and J.H. Grove. (1990). Testing soil for phosphorus. In *Soil Testing and Plant Analysis, 3rd edition*, ed. R.L. Westermann. Madison, Wisconsin, USA: Soil Science Society of America, pp. 141-179.

Fox, R.L. (1979). Comparative response of field grown crops to phosphate concentrations in soil solutions. In *Stress Physiology in Crop Plants*, eds. H. Mussel and R. Stapes. New York, USA: John Wiley & Sons, pp. 81-106.

Fox, R.L. (1981). External phosphorus requirements of crops. In *Chemistry in the Soil Environment*, eds. R.H. Dowdy, V.V. Vok, and D.E. Baker. Madison, Wisconsin, USA: American Society of Agronomy, Soil Science Society of America, pp. 223-239.

Fox, R.L. (1988). External P requirement of plants and their nutrition from fertilizer and soil P. In *Proceedings of the Phosphorus Symposium*. Pretoria, South Africa: Soils and Irrigation Research Institute, pp. 112-119.

Gardner, W.K., D.G. Parbery, and D.A. Barber. (1981). Proteoid root morphology and function in *Lupinus albus*. *Plant and Soil* 60: 143-147.

Gardner, W.K., D.G. Parbery, and D.A. Barber. (1982a). The acquisition of phosphorus by *Lupinus albus* L. I. Some characteristics of the root/soil interface. *Plant and Soil* 69: 19-32.

Gardner, W.K., D.G. Parbery, and D.A. Barber. (1982b). The acquisition of phosphorus by *Lupinus albus* L. II. The effect of varying phosphorus supply and soil type on some characteristics of the soil/root interface. *Plant and Soil* 68: 33-41.

Gerdemann, J.W. (1968). Vesicular-arbuscular mycorrhizae and plant growth. *Annual Review of Phytopathology* 6: 394-418.

Gerdemann, J.W. (1975). Vesicular-arbuscular mycorrhizae. In *The Development and Function of Roots, 3rd Cabot Symposium, Harvard Forest, April 1974*, eds J.G. Torrey and D.T. Clarkson. London, UK: Academic Press, pp. 575-591.

Gilkes, R.J. (1975). Some properties of granulated superphosphate and its behaviour in the soil. *Australian Journal of Soil Research* 13: 203-215.

Gilkes, R.J. and M.D.A. Bolland. (1990). The poor performance of rock phosphate fertilizers in Western Australia: Part 2. The chemical explanations. *Agricultural Science* 3: 44-47.

Gilkes, R.J. and R. Lim-Nunez. (1980). Poorly soluble phosphates in Australian superphosphates: their nature and availability to plants. *Australian Journal of Soil Research* 31: 85-95.

Gilkes, R.J. and P. Mangano. (1983). Poorly soluble iron-aluminium phosphates in ammonium phosphate fertilizers: their nature and availability to plants. *Australian Journal of Soil Research* 21: 183-194.

Ham, G.E., W.W. Nelson, S.D. Evans, and R.D. Frazier. (1973). Influence of fertilizer placement on yield response of soybeans. *Agronomy Journal* 65: 81-84.

Hammond, L.L. (1978). Agronomic measurements of phosphate rock effectiveness. *Seminar on Phosphate Rock for Direct Application, Haifa, Israel.* Muscle Shoals, Alabama, USA: Special Publication IFDC-SI, International Fertilizer Development Centre (IFDC), pp. 147-173.

Hammond, L.L., S.H. Chien, and A.U. Mokwunye. (1986). Agronomic value of unacidulated and partially acidulated phosphate rocks indigenous to the tropics. *Advances in Agronomy* 40: 89-140.

Haynes, R.J. (1992). Relative ability of a range of crop species to use phosphate rock and monocalcium phosphate as P sources when grown in soil. *Journal of Science of Food and Agriculture* 60: 205-211.

Helyar, K.R. (1978). Effects of aluminium and manganese toxicities on legume growth. In *Mineral Nutrition of Legumes in Tropical and Sub-Tropical Soils*, eds. C.S. Andrew and E.J. Kamprath. Melbourne, Australia: CSIRO Publishing, pp. 207-231.

Hingston, F.J., A.M. Posner, and J.P. Quirk. (1972). Anion adsorption by goethite and gibbsite. 1. The role of the proton in determining adsorption envelopes. *Journal of Soil Science* 23: 177-192.

Hingston, F.J., R.J. Atkinson, A.M. Posner, and J.P. Quirk. (1968). Specific adsorption of anions. *Nature* 215: 1459-1461.

Hinsinger, P. and R.J. Gilkes. (1995). Root-induced dissolution of phosphate rock in the rhizosphere of lupins grown in alkaline soil. *Australian Journal of Soil Research* 33: 477-489.

Hoffland, E., G.R. Findenegg, and J.A. Nelemans. (1989a). Solubilization of rock phosphates by rape. I. Evaluation of the role of the nutrient uptake pattern. *Plant and Soil* 113: 155-160.

Hoffland, E., G.R. Findenegg, and J.A. Nelemans. (1989b). Solubilization of rock phosphate by rape. II. Local root exudation of organic acids as a response to P starvation. *Plant and Soil* 113: 161-165.

Holford, I.C.R. and G.E.C. Mattingly. (1975). The high and low-energy phosphate adsorbing surfaces in calcareous soils. *Journal of Soil Science* 26: 407-417.

Holford, I.C.R., R.W.M. Wedderburn, and G.E.C. Mattingly. (1974). A Langmuir two-surface equation as a model for phosphate adsorption by soils. *Journal of Soil Science* 25: 242-254.

Hughes, J.C. and R.J. Gilkes. (1994). Rock phosphate dissolution and bicarbonate-soluble P in some soils from south-western Australia. *Australian Journal of Soil Research* 32: 767-779.

Jarvis, R.J. and M.D.A. Bolland. (1991). Lupin grain yields and fertilizer effectiveness are increased by banding superphosphate below the seed. *Australian Journal of Experimental Agriculture* 31: 357-366.

Kanabo, I.A.K. and R.J. Gilkes. (1988). The effects of moisture regime and incubation on the dissolution of North Carolina phosphate rock in soil. *Australian Journal of Soil Research* 26: 153-163.

Khasawneh, F.E. and E.C. Doll. (1978). The use of phosphate rock for direct application to soils. *Advances in Agronomy* 30: 159-206.

Kitterick, J.A. and Jackson M.L. (1955). Rate of phosphate reaction with soil minerals and electron microscope observations on the reaction mechanism. *Soil Science Society of America Proceedings* 19: 292-295.

Kitterick, J.A. and M.L. Jackson. (1956). Electron microscope observations of the reactions of phosphate with minerals, leading to a unified theory of phosphate fixation in soils. *Journal of Soil Science* 7: 81-88.

Kolaian, J.H. and A.J. Ohlrogge. (1959). Principles of nutrient uptake from fertilizer bands: IV. Accumulation of water around the bands. *Agronomy Journal* 51: 106-108.

Kumar, V., R.J. Gilkes, and M.D.A. Bolland. (1992). The residual value of rock phosphate and superphosphate from field sites assessed by glasshouse bioassay using three plant species with different external P requirements. *Fertilizer Research* 30: 31-38.

Lawton, K. and J.A. Vomocil. (1954). The dissolution and migration of phosphorus from granular superphosphate in some Michigan soils. *Soil Science Society of America Proceedings* 18: 26-32.

Lehr, J.R. and W.E. Brown. (1958). Calcium phosphate fertilizers. II. A petrographic study of their alteration in soils. *Soil Science Society of America Proceedings* 22: 29-32.

Lehr, J.R., W.E. Brown, and E.H. Brown. (1959). Chemical behaviour of monocalcium phosphate monohydrate in soils. *Soil Science Society of America Proceedings* 23: 3-7.

Lehr, J.R., E.M. Brown, A.W. Frazier, J.P. Smith, and R.D. Throsher. (1967). *Crystallographic Properties of Fertilizer Compounds.* Muscle Shoals, Alabama, USA: National Fertilizer Development Centre.

Leon, L.A., W.E. Fenster, and L.L. Hammond. (1986). Agronomic potential of eleven phosphate rocks from Brazil, Colombia, Peru, and Venezuela. *Soil Science Society of America Journal* 50: 798-802.

Lewis, D.G. and J.P. Quirk. (1965). Diffusion of phosphate to plant roots. *Plant Nutrient Supply and Movement*, Technical Report Series No. 48, International Atomic Energy Agency, Vienna, Austria, pp. 71-77.

Lindsay, W.L. and H.F. Stephenson. (1959a). Nature of the reactions of monocalcium phosphate monohydrate in soils. I. The solution that reacts with the soil. *Soil Science Society of America Proceedings* 23: 12-18.

Lindsay, W.L. and H.F. Stephenson. (1959b). Nature of the reactions of monocalcium phosphate monohydrate in soils. II. Dissolution and precipitation reactions involving iron, aluminum, manganese, and calcium. *Soil Science Society of America Proceedings* 23: 18-22.

Low, P.F. and C.A. Black. (1948). Phosphate induced decomposition of kaolinite. *Soil Science Society of America Proceedings* 12: 180-184.

Marschner, H., V. Römheld, and I. Cakmak. (1987). Nutrient availability in the rhizosphere. *Journal of Plant Nutrition* 10: 1178-1184.

McClean, E.O. and R.W. Wheeler. (1964). Partially acidulated rock phosphate as a source of phosphorus to plants. I. Growth chamber studies. *Soil Science Society of America Proceedings* 28: 545-550.

McClean, E.O., R.W. Wheeler, and J.D. Watson. (1965). Partially acidulated rock phosphate as a source of phosphorus to plants. II. Growth chamber and field corn studies. *Soil Science Society of America Proceedings* 29: 625-628.

Mengel, K. and E.A. Kirkby. (1987). *Principles of Plant Nutrition.* Berne, Switzerland: International Potash Institute.

Mosse, B. (1973). Advances in the study of vesicular-arbuscular mycorrhiza. *Annual Review of Phytopathology* 11: 171-196.

Muljadi, D., A.M. Posner, and J.P. Quirk. (1966a). The mechanism of phosphate adsorption by kaolinite, gibbsite and pseudoboemite. Part I. The isotherm and the effect of pH on adsorption. *Journal of Soil Science* 17: 212-229.

Muljadi, D., A.M. Posner, and J.P. Quirk. (1966b). The mechanism of phosphate adsorption by kaolinite, gibbsite and pseudoboemite. Part II. The location of the adsorption sites. *Journal of Soil Science* 17: 230-237.

Muljadi, D., A.M. Posner, and J.P. Quirk. (1966c). The mechanism of phosphate adsorption by kaolinite, gibbsite and pseudoboemite. Part III. The effect of temperature on the adsorption. *Journal of Soil Science* 17: 238-247.

Norrish, K. and H. Rosser. (1983). Mineral phosphate. In *Soils: An Australian Viewpoint*, eds. CSIRO Division of Soils. Melbourne, Australia: CSIRO Publishing/Academic Press, pp. 335-361.

Nishimoto, R.K., R.L. Fox, and P.E. Parvin. (1977). Responses of vegetable crops to phosphorus concentrations in soil solution. *Journal of American Society of Horticultural Science* 102: 705-709.

Ozanne, P.G. and T.C. Shaw. (1967). Phosphate sorption by soils as a measure of the phosphate requirements of pasture growth. *Australian Journal of Agricultural Research* 18: 601-612.

Pairunan, A.K., A.D. Robson, and L.K. Abbott. (1980). The effectiveness of vesicular-arbuscular mycorrhizas in increasing growth and phosphorus uptake of subterranean clover from phosphorus sources of different solubilities. *The New Phytologist* 84: 327-338.

Rajan, S.S.S., K.W. Perrott, and W.M.H. Saunders. (1974). Identification of phosphate-reactive sites of hydrous alumina from proton consumption during phosphate adsorption at constant pH values. *Journal of Soil Science* 25: 438-447.

Robinson, N. (1980). Phosphoric acid technology. In *The Role of Phosphorus in Agriculture*, eds. F.E. Khasawneh, E.C. Sample, and E.J. Kamprath. Madison, Wisconsin, USA: American Society of Agronomy, pp. 151-193.

Rudd, C.L. and N.J. Barrow. (1973). The effectiveness of several methods of applying superphosphate on yield response by wheat. *Australian Journal of Experimental Agriculture and Animal Husbandry* 13: 430-433.

Sale, P.W.G. and G.J. Blair. (1989). Low solubility phosphate fertilizers for pastures: an alternative prospect. *Agricultural Science* 2: 34-39.

Sample, E.C., R.J. Soper, and G.J. Racz. (1980). Reactions of phosphate fertilizers in soils. In *The Role of Phosphorus in Agriculture*, eds. F.E. Khasawneh, E.C. Sample, and E.J. Kamprath. Madison, Wisconsin, USA: American Society of Agronomy, pp. 263-310.

Sommer, A.L. (1936). The relationship of the phosphate concentration of soil solution cultures to the type and size of root systems and the time of maturity of certain plants. *Journal of Agricultural Research* 52: 133-148.

Strong, J. and G.J. Racz. (1970). Reactions products of applied orthophosphate in some Manitoba soils as affected by soil calcium and magnesium content and time of incubation. *Soil Science* 110 : 258-262.

Thomas, G. W. and D.E. Peaslee. (1973). Testing soils for phosphorus. In *Soil Testing and Plant Analysis*, ed. L.M. Walsh and J.O. Beaton. Madison, Wisconsin, USA: Soil Science Society of America, pp. 115-132.

Trumble, H.C. and C.M. Donald. (1938). The relation of phosphate to the development of seeded pasture on a podsolized sand. CSIR, Melbourne, Australia. CSIR Bulletin No. 116.

Waggaman, W.H. (1969). *Phosphoric Acid, Phosphates, and Phosphatic Fertilizers.* 2nd ed. New York, USA: Hafner Publication Company.

Welch, L.F., D.L. Mulvaney, L.V. Boone, G.E. McKibben, and J.W. Pendelton. (1966). Relative efficiency of broadcast versus banded phosphorus for corn. *Agronomy Journal* 58: 283-287.

White, M.S. (1974). The liquid phase of superphosphate. *New Zealand Journal of Science* 17: 171-182.

Wild, A. (1950). The retention of phosphate by soil. A review. *Journal of Soil Science* 1: 221-238.

Young, D.R. and C.H. Davis. (1980). Phosphate fertilizers and process technology. In *The Role of Phosphorus in Agriculture*, eds. F.E. Khasawneh, E.C. Sample, and E.J. Kamprath. Madison, Wisconsin, USA: American Society of Agronomy, pp. 195-226.

SUBMITTED: 08/20/97
ACCEPTED: 11/11/97

Micronutrient Fertilizers

Larry M. Shuman

SUMMARY. Micronutrient fertilizer sources are mostly sulfates for Zn, Cu, and Mn, but chelates are the usual Fe source, and borax and sodium molybdate are used for B and Mo, respectively. Soil pH is the soil property that most influences micronutrient availability, and for all but Mo, the higher the soil pH, the lower is the plant availability. For Mo, liming can actually prevent deficiencies. Other soil properties that are important in bioavailability are organic matter content, especially for Cu, oxidation/reduction conditions, especially for Fe and Mn, soil texture, Fe and Al oxide content and soil moisture conditions. *[Article copies available for a fee from The Haworth Document Delivery Service: 1-800-342-9678. E-mail address: getinfo@haworthpressinc.com]*

KEYWORDS. Boron, copper, iron, manganese, micronutrients, molybdenum, zinc

INTRODUCTION

There are many soil properties which influence micronutrient availability, including soil type, texture, organic matter content, oxide content, moisture, and oxidation/reduction conditions, along with the

Larry M. Shuman, Professor of Soil Chemistry, University of Georgia, Crop and Soil Sciences Department, Georgia Experiment Station, Griffin, GA 30223-1797 USA (E-mail: lshuman@gaes.griffin.peachnet.edu).

The author's research is supported by State and HATCH funds allocated to the Georgia Experiment Station.

[Haworth co-indexing entry note]: "Micronutrient Fertilizers." Shuman, Larry M. Co-published simultaneously in *Journal of Crop Production* (Food Products Press, an imprint of The Haworth Press, Inc.) Vol. 1, No. 2 (#2), 1998, pp. 165-195; and: *Nutrient Use in Crop Production* (ed: Zdenko Rengel) Food Products Press, an imprint of The Haworth Press, Inc., 1998, pp. 165-195. Single or multiple copies of this article are available for a fee from The Haworth Document Delivery Service [1-800-342-9678, 9:00 a.m. - 5:00 p.m. (EST). E-mail address: getinfo@haworthpressinc.com].

most obvious being soluble micronutrient content. In reviewing literature on micronutrient fertilizers, it is impossible to ignore soil factors that influence the effectiveness of the various sources available to the grower. Thus, this chapter not only gives information about fertilizers, but also about the soil properties which influence the availability of the elements to plants and, in some cases about vulnerability to leaching. The chapter will not deal with soil or plant critical levels or effects of the plant root rhizosphere on availability. The literature covered is primarily from the last 20 years, so some informative earlier literature is omitted. The reader may notice that the space given to Zn is much greater than that for the other elements. That is simply because the literature is more extensive for Zn than for the others covered, namely, Cu, Mn, Fe, B, and Mo. The minor micronutrients, which are essential to plants, such as Cl and Ni are not covered because they are not added as fertilizer elements.

ZINC

Sources

Zinc is usually added as an inorganic fertilizer, especially in the form of $ZnSO_4$. Other forms added include organics, such as lignosulfonate and synthetic chelates (Table 1). The popularity of $ZnSO_4$ stems from its high solubility, low cost, and wide availability (Martens and Westermann, 1991). Other sources that are being utilized more today than in the recent past are sewage sludges and animal manures. The variability of the concentrations of Zn, especially in municipal sewage sludges, makes it imperative that applications be based on analysis of the materials, so as not to exceed loading rates of Zn or other metals that may be contained in the sludges.

Source comparisons for Zn in the current literature are not as prevalent as in decades past, but now are aimed more at providing adequate Zn to the crop while not endangering the environment. For corn, Gallagher, Murphy, and Ellis (1978) found that organic Zn sources (ZnEDTA and Zn-Reax, a lignosulfonate) tended to supply more Zn to the plants than inorganic sources (ZnO and $ZnSO_4$) on an acidic soil under low temperatures, but the opposite was true for higher temperatures. Organic Zn sources were slightly more effective than inorganic forms on calcareous soils. The synthetic chelate ZnEDTA was found

TABLE 1. Zinc fertilizer sources[+]

Source	Formula	Concentration of Zn, g kg^{-1}[‡]
Zinc ammonia complex	Zn-NH$_3$	100
Zinc carbonate	ZnCO$_3$	520-560
Zinc chelate	Na$_2$ZnEDTA	140
Zinc chelate	NaZnHEDTA	90
Zinc chelate	NaZnNTA	90
Zinc chloride	ZnCl$_2$	480-500
Zinc frits	Fritted glass	100-300
Zinc lignosulfonate		50-80
Zinc nitrate	Zn(NO$_3$)$_2$·6H$_2$O	220
Zinc oxide	ZnO	500-800
Zinc oxysulfate	ZnO + ZnSO$_4$	400-550
Zinc polyflavonoid		50-100
Zinc sulfate monohydrate	ZnSO$_4$·H$_2$O	360
Zinc sulfate heptahydrate	ZnSO$_4$·7H$_2$O	230
Basic zinc sulfate	ZnSO$_4$·4Zn(OH)$_2$	550

[+]Data from Martens and Westermann (1991).
[‡]Approximate concentration.

to be a more effective source for corn than Zn lignosulfonate, and resin coating the Zn-EDTA improved its performance (Alvarez, Obrador, and Rico, 1996). Moraghan (1996) found that ZnSO$_4$ and ZnEDTA gave the same increases in concentrations of Zn in the seed of navy beans. Banding the same sources gave lower seed Zn concentrations than when incorporated throughout the soil.

Soil samples taken from field plots amended with three Zn sources were extracted successively as saturation extracts. After six extractions, Zn in the extracts was below detection limits indicating that Zn fertilizers do not dissolve quickly to bring their concentrations back to the original levels in soils. This slow release tends to decrease potential leaching losses, but also may limit the resupply of Zn to the plant (Gammon, 1976). Zinc movement in the soil profile was greater with ZnEDTA than with ZnSO$_4$, showing that the synthetic chelate keeps the Zn in the soil solution and makes it more vulnerable to leaching

(Raghbir, Shukla, and Singh, 1976). The same experiment showed that urea, monocalcium phosphate, calcium carbonate, and farmyard manure all decreased the movement of Zn in columns of a loamy sand and a loam. Thus, not only Zn fertilizer sources, but also other soil amendments may have a significant influence on Zn availability and mobility.

Bioavailability

Soil pH

Zinc availability to plants from fertilizers and other sources depends to a great extent on soil pH which governs the solubility of Zn solid phases and greatly influences Zn adsorption and release by soil colloidal surfaces. There are many published studies which indicate that both Zn uptake by plants and Zn extractability in soils is decreased by increasing soil pH. Wu and Aasen (1994) concluded that Zn availability was most affected above pH 6.5 but below that pH, the availability was determined more by soil Zn quantity than pH. This finding is supported by data for rice where a pH decrease from 8.0 to 6.5 resulted in a linear decrease in Zn extraction level in soils (Sedberry et al., 1980). Several studies have shown that plant Zn concentration is better predicted by including both extractable soil Zn and soil pH in a regression equation as opposed to including only extractable soil Zn (Peaslee, 1980; Lins and Cox, 1988; Cox, 1990; Parker et al., 1990). In a field experiment with corn it was shown that increasingly more Zn was required to increase corn shoot and ear leaf Zn concentration as soil pH increased (Boswell, Parker, and Gaines, 1989). Similar findings were reported for peanuts in that an increase in extractable soil Zn increased leaf Zn more at a soil pH of 4.6 than at pH 6.6 (Davis-Carter, Parker, and Gaines, 1991). In calcareous soils, the pH and Zn availability relationship can be opposite to that in acid soils. Gough, McNiel, and Severson (1980) found that for the A-horizon of calcareous soils there was a positive relation between soil pH and extractable Zn concentrations that was apparently related to the soil levels of Zn carbonate.

Other Soil Property Effects

Soil Zn availability has often been found to be linked positively with soil organic matter. Jahiruddin et al. (1992) showed that the influence of soil pH on Zn availability in Scottish soils was more evident for soils low in organic matter (< 6.5%) than in soils with high organic matter contents (> 6.5%). Zinc availability is usually higher in finer-textured soils with higher clay contents and with higher organic matter contents (MacLean and Langille, 1983; Brennan and Gartrell, 1990; Sharma, Sidhu, and Nayyar, 1992).

Soils with increasing $CaCO_3$ content show a decrease in Zn availability to plants (Van Breemen, Quijano, and Le-Ngoc, 1980). In a toposequence it was found that no Zn deficiency occurred at the highest elevation, and progressively severe Zn deficiencies occurred in successively lower fields. The fields at the lower elevation showed an increase in organic matter and free carbonate content (Van Breemen, Quijano, and Le-Ngoc, 1980).

Soil Fe oxide concentrations can affect Zn availability, since Fe oxides adsorb Zn at specific sites (McBride, 1994). However, it was shown that organic matter is still more important than Fe and Mn oxides in influencing soil Zn availability, especially at the toxic end of the spectrum (White and Chaney, 1980). Conventional tillage operations such as disk harrowing can increase soil pH and thereby decrease Zn availability as compared to non-tilled land (Wood and White, 1986). With an increase in pH, Zn uptake decreased sharply in rice plants under aerobic conditions but much less so under anaerobic conditions (Jugsujinda and Patrick, 1977).

Soil Phosphorus

There is an interaction between P and Zn availability, which is usually attributed to plant effects, but has also been reported to be a soil process. Zinc has been shown to interact with soil P in strawberry plants (May and Pritts, 1993), being significant, along with soil pH, in multiple regression expressions to predict Zn uptake (Peaslee, 1980; Shang and Bates, 1987). For soybeans on a Southeastern Ultisol, Zn deficiency was induced by high P rates and high pH, and the magnitude of the yield decrease corresponded well with the intensity of deficiency symptoms (Adams, Adams, and Odom, 1982). Corn yields increased in response to Zn application on soils of Zimbabwe which were either limed or had received 240 kg P_2O_5 ha^{-1} (Tagwira, Phia, and Mugwira, 1993). The increase in soil CEC caused by P application correlated with the decrease in plant tissue Zn concentration. In saline alluvial Haplaquepts, available P_2O_5 was correlated negatively with available Zn (Maji, Chatterji, and Bandyopadhyay, 1993).

Soil Reactions

Diffusion

Diffusion of Zn from the fertilizer granule and subsequently to the plant root is a major mechanism influencing the availability of Zn to plants. Soils high in clay content, high in organic matter content and with high cation exchange capacity exhibit slow diffusion of Zn (Modaihsh, 1990). As indicated above, the plant availability of Zn increases with increases in soil organic matter content. Soils with higher organic matter have more total and more extractable Zn (Shuman, 1988) leading to less dependence on diffusion to supply Zn to the root. Modaihsh (1990) found that organic matter amended at a rate of 5% decreased the diffusion and extractability of $ZnSO_4$ in both calcareous and noncalcareous soils. However, the application of elemental S to decrease pH, increased Zn diffusion from $ZnSO_4$ fertilizer in noncalcareous soils, while there was little effect in calcareous soils. Application of manures to soils of varying textures caused more decreases in Zn concentration at the root surface with respect to the bulk solution than where no manure was applied. This result corresponded with increases in measured Zn uptake by plants with manure application (Sharma and Deb, 1991). In this case soluble organics may have increased the diffusion and availability to the wheat plants. Cropping corn on a calcareous soil decreased the concentrations of Zn in the exchangeable fraction due to plant root processes (Ma and Uren, 1996). These authors indicated that Zn diffused into microporous solids, which was the rate-limiting reaction.

Forms of Zinc

Amending soils with Zn as a salt will cause an imbalance in the equilibrium of soil Zn, since the salt is much more soluble than equilibrium solid forms such as the carbonate, oxide, or various hydroxides. The reactions that take place cause the fertilizer Zn to assume less and less plant available forms with time. In a Canadian study, soil-applied Zn became preferentially associated with the hydrous oxide fractions and the functional groups of soil organic matter (Liang, Karamanos, and Stewart, 1991). This report is substantiated by Iwassaki, Yoshikawa, and Sakurai (1993) who found that added fertilizer Zn in a greenhouse trial was specifically adsorbed or occluded by the oxides and hydroxides of soil Fe and Mn. They found a negative correlation between exchangeable Zn concentrations and soil pH. Haynes and Swift (1985) found that concentrations of the exchangeable, organic matter bound, and oxide forms of Zn increased as the soil pH was decreased from 7.0 to 3.8 using sulphuric acid additions. Adding $ZnSO_4$ to southeastern Coastal Plain soils increased the soluble forms of Zn (exchangeable and organic matter bound) whereas liming led to the redistribution of Zn into less soluble fractions (oxide) (Davis-Carter and Shuman, 1993). Zinc in the exchangeable fraction was best correlated with the soil CEC, and plant Zn concentration was most highly correlated with the exchangeable form. Shuman (1988) showed that adding organic material to soils increased the concentration of Zn in the Mn and Fe oxide fractions at the expense of the other fractions. Released Zn may have been occluded in these fractions or strongly adsorbed. Soil-amended Zn was evident in the exchangeable and organic matter fractions. Adding the synthetic organic chelator EDTA to soils redistributed Zn from the organic matter fraction to the exchangeable fraction making it more vulnerable to leaching (Li and Shuman, 1996). Increasing the redox potential of soils (oxidation) increased the dissolved and exchangeable Zn concentrations and greatly increased plant Zn concentrations (Gambrell and Patrick, 1989). Thus, the above research indicates that for acidic soils, fertilizer Zn associates mainly with the exchangeable and organic matter fractions initially, but reverts to less available forms such as the oxide and residual fractions with time.

Residual Zinc

Zinc fertilizers and organic sources may suffice for more than one season or for more than one crop in multiple cropping schemes. On the U.S. Coastal Plain, Zn accumulated in the soil over time, and eventually more Zn became available from residual Zn than from recently applied Zn, especially at the highest pH (Boswell, Parker, and Gaines, 1989). In a field trial in India, Zn sources as farmyard manure (FYM), $ZnSO_4$, and zincated urea were added to a sorghum crop and the effectiveness on a subsequent wheat crop tested (Indulkar and Malewar, 1994). All sources including FYM increased the yield of the following wheat crop. In a greenhouse pot experiment with corn, the second crop contained elevated Zn concentrations from fertilization of the first crop, but there was no increased plant growth for the second crop (Karimian and Yasrebi, 1995).

COPPER

Sources

Copper is needed by plants in lower quantities than Zn and is generally found at lower concentrations in soils than Zn. However, the reactions of Cu in the soil and the sources of Cu used as fertilizers are often similar to Zn because of the similarity of the elements being divalent metal cations. Sources of Cu commonly used as fertilizers are shown in Table 2. As for Zn, the most widely used form of fertilizer Cu is the sulfate form, because it is very soluble, is relatively inexpensive and has widespread availability (Martens and Westermann, 1991). Besides inorganic sources, sewage sludges and manures (especially pig [Reed et al., 1993]) are used as sources of Cu.

TABLE 2. Copper fertilizer sources[†]

Source	Formula	Concentration of Cu, g kg^{-1}[‡]
Basic copper sulfates	$CuSO_4 \cdot 3Cu(OH)_2$	130-530
Copper chelate	$Na_2CuEDTA$	130
Copper chelate	NaCuHEDTA	90
Copper chloride	$CuCl_2$	470
Copper lignosulfonate		50-80
Copper polyflavonoid		50-70
Copper sulfate monohydrate	$CuSO_4 \cdot H_2O$	350
Copper sulfate pentahydrate	$CuSO_4 \cdot 5H_2O$	250
Cuprous oxide	Cu_2O	890
Cupric oxide	CuO	750

[†]Data from Martens and Westermann (1991).
[‡]Approximate concentration.

Care must be taken to monitor loading rates of these materials in order to avoid Cu toxicity and pollution of the soil as well as surface and groundwaters.

Copper deficiencies are usually corrected by soil application of Cu inorganic fertilizers as opposed to foliar applications. In an oat experiment where 190 mg Cu kg^{-1} was added as $CuSO_4$, $CuCl_2$, CuO, or Cu-EDTA, the EDTA source of Cu was most available (Jurkowska and Rogoz, 1977). Adding Cu as a constituent of sewage sludge or as an inorganic salt increased the plant content of wheat, but raising the soil pH decreased the tissue concentrations regardless of the source (Tadesse et al., 1991). Additions of poultry litter to pastures over a period of 15 to 28 years increased the Cu concentrations of plant tissue compared to a control where no litter was applied (Kingery et al., 1993). Adding Cu-rich pig manure at 365 kg Cu ha^{-1} (well in excess of the USEPA maximum safe-loading rate) had no effect on corn grain yields or Cu concentration (Reed et al., 1993). The applied Cu was evidently held in largely unavailable forms.

Bioavailability

Soil pH

As is the case for Zn, Cu availability and extractability of soil Cu decreases with increasing soil pH (MacLean and Langille, 1983; Haynes and Swift, 1985). Liming soils from pH 3.9 to 6.7 significantly decreased the plant Cu in highbush blueberry, but did not affect dry matter yields (Haynes and Swift, 1985). In a sandy soil in Florida, Cu contamination was less injurious to citrus seedlings at pH 6 than at pH 7 indicating that Cu was less available at the higher pH (Mozaffari, Alva, and Chen, 1996). In a calcareous soil the plant Cu concentrations were positively correlated with A-horizon soil pH, presumably due to higher levels of soil metal carbonates increasing availability (Gough, McNeil, and Severson, 1980).

Other Soil Property Effects

Copper is closely associated with organic matter in soils and is the most tightly bound in organic complexes of any of the divalent transition metals (McBride, 1994). The retention of Cu by soil colloids has been ranked in the order organics > Fe/Mn oxides >> layer silicates (Baker and Sneft, 1995). Thus, it has been found that extractable soil Cu is usually negatively correlated with soil organic matter content (MacLean and Langille, 1983). In arid-zone soils, the opposite relation was found where DTPA-extractable soil Cu increased with an increase in organic carbon content (Sharma, Sidhu, and Nayyar, 1992). Petruzzelli, Guidi, and Lubrano (1978) showed that Cu adsorption was greatly decreased by removing the organic matter. Soil clay content, with the concurrent influence on CEC, usually shows a positive correlation with extractable soil Cu (Sharma, Sidhu, and Nayyar, 1992).

Soil Reactions

Copper added as fertilizer is in soluble forms, but soon reverts to other forms including adsorbed and associated with organic matter and oxides. In a study on the forms of Cu from amendments in 55 Canadian organic soils, only 0.85% of the total was water-soluble and 0.67% was on the exchange complex (Levesque and Mathur, 1986). That study showed most of the Cu to be in residual forms that were unextractable (possibly of natural origin), while the next most abundant fraction was strongly complexed (28%). In another study on organic soils it was found that the immediately available Cu can be replenished by the oxidation of sulfide-associated Cu (Mathur and Levesque, 1989).

Residual Copper

Copper fertilizers are known to have residual effects mainly because Cu is not mobile in soils and is usually added in excess of immediate crop requirements. However, fertilizer Cu does decrease in effectiveness with time as found by Levesque and Mathur (1986) for organic soils where residual fertilizer value decreased as its residence time in the soil increased. Copper added to four Ultisols and one Inceptisol at a rate of 4.48 kg ha^{-1} was found to increase extractable Cu by two standard deviations for an average of 16 years (Cox, 1992). In Australia, the residual effects of Cu applications for a period of 18-21 years at recommended rates for wheat were predicted to last from 20 to 30 years depending on the soil type (Brennan, 1994).

McLaren and Ritchie (1993) found that a large portion of applied Cu to a lateritic sandy soil was redistributed into the residual fraction over a course of 20 years for plots with lower rates of Cu, but for higher rates a much smaller proportion was found in the residual form.

MANGANESE

Sources

Manganese is most often applied as a sulfate salt, but other inorganic and organic sources are available for use. Of the inorganic sources shown in Table 3 the one most used after the sulfate is finely ground MnO (Martens and Westermann, 1991). Foliar sprays are used when deficiencies are evident early in the growing season and here the EDTA form is often applied (Ohki et al., 1987). Broadcasting Mn can result in the soluble form reverting to oxide forms that are unavailable to plants. Therefore, banding is often recommended.

A comparison of Mn sources for Bermudagrass on a sandy soil showed that the sulfate and chelate sources provided short-term correction, but little response was obtained with a frit source (Snyder, Burt, and Gascho, 1979). For peanuts, Mn gave responses on a sandy soil limed to pH near 6.8 at rates up to 40 kg ha^{-1} (Parker and Walker,

TABLE 3. Manganese fertilizer sources[+]

Source	Formula	Concentration of Mn, g kg^{-1}[‡]
Manganese chelate	Na$_2$MnEDTA	50-120
Manganese chloride	MnCl$_2$	170
Manganese frits	Fritted glass	100-350
Manganese lignosulfonate		50
Manganous oxide	MnO	400
Manganese oxysulfate		280
Manganese polyflavonoid		50-70
Manganese sulfate	MnSO$_4$	230-280

[+]Data from Martens and Westermann (1991).
[‡]Approximate concentration.

1986; Parker et al., 1988). Alley (1975) found that Mn deficiency in soybeans was corrected best by foliar application of 2.2 kg ha^{-1}, but broadcast Mn at 44.8 kg ha^{-1} was satisfactory. Banding 16.8 kg ha^{-1} with the seed caused stand reductions under low soil moisture conditions. In a field trial with soybeans on a sandy Coastal Plain soil, $MnSO_4$, MnO and fritted Mn were all found to be satisfactory sources for yield increases, but a soil-applied chelate source did not increase yields above the control (Shuman et al., 1979) (Figure 1). Leaf and soil concentrations of Mn were increased by the fertilizer sources in the order $MnSO_4$ > MnO > fritted Mn > Mn chelate.

Bioavailability

Soil pH

Manganese extractability and plant availability is highly pH-sensitive; much research has been reported showing that Mn availability decreases as soil pH is increased. Manganese fertilization is often not needed for soils with pH values lower than about 5.7 (Alley et al., 1978; Parker and Walker, 1986). Manganese deficiencies and responses to Mn fertilization usually occur where the total soil Mn is relatively low and soil pH is high, often from overliming legumes such as soybeans and peanuts (Shuman et al., 1979; Parker and Walker, 1986). In calcareous soils, Mn availability can be positively related to soil pH which is a result of Mn carbonate levels (Gough, McNeil, and Severson, 1980).

Manganese availability is sometimes increased in organic or calcareous soils by acidification using elemental sulphur or acid-forming nitrogenous fertilizers. Remon et al. (1977) found sulphur application successful in increasing Mn supply to Bermudagrass. Band application of S along with Mn fertilizer was the most cost-effective way to remediate Mn deficiency in celery grown on a Florida Histosol (Beverly, 1987). Snyder, Burt, and Gascho (1979) added ammonium sulphate to increase the Mn availability to Bermudagrass. Soil solution Mn concentration was significantly increased by band placement of $(NH_4)_2SO_4$, NH_4Cl, and urea (Petrie and Jackson, 1984).

Other Soil Property Effects

Although soil pH is a major factor in Mn reactions and availability in soils, other soil properties and processes (such as soil solution

FIGURE 1. Soybean yield and leaf Mn concentration for four Mn rates and three Mn fertilizer sources. Values within a Mn source with the same letter are not different at the 5% level according to Duncan's Multiple Range Test. The unit q/ha is equal to 100 kg/ha. (Adapted from Shuman et al., 1979).

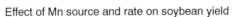

Effect of Mn source and rate on soybean yield

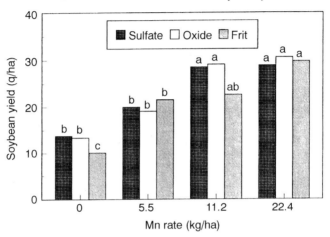

Effect of Mn source and rate on soybean leaf Mn

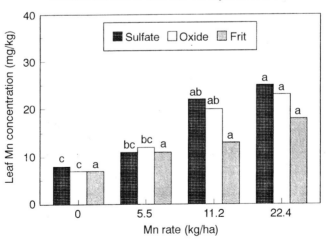

conductivity, cation exchange capacity and organic carbon, clay, and $CaCO_3$ contents) also play a part (Van Eysinga, Van Dijk, and De Bes, 1978; Sangwan and Kuldeep, 1993; Ahangar et al., 1995). Adding organic materials to soils affects Mn availability through several processes, including changes in pH, changes in oxidation/reduction conditions, and increasing concentrations of soluble complexing ligands. Addition of decomposing ryegrass or alfalfa residues to flooded soils increased the extractable Mn, presumably due to decreasing redox potentials (Velez-Ramos and Standifer, 1977). Amending soils with sewage sludge increased the amounts of organic ligands such as tartaric, pyruvic, and protocatechuic acids, in the soil solution (Hue, 1988). The soluble Mn in soil solution was found to be 76 to 99% complexed by these soluble organic ligands accounting for the high, phytotoxic levels of Mn in the soil solutions.

Finally, soil P levels can influence Mn availability, due to both soil chemical reactions and plant effects. Boyle and Lindsay (1986) showed that Mn phosphate can control Mn solubility under certain soil conditions, so P levels in soil can have an influence on Mn availability. Neilsen et al. (1992) suggested that high soil P resulted in elevated plant P concentration that interfered with the uptake and/or translocation of Mn in wheat.

IRON

Sources

Deficiency of Fe in plants is restricted mostly to areas of alkaline and calcareous soils. Acidic soils usually supply sufficient Fe for plant growth due to the acid solubility of Fe compounds in soils. Supplying Fe to plants by inorganic fertilizers is difficult because the soluble Fe quickly reverts to plant-unavailable forms. Therefore, chelated forms are the most effective, albeit quite expensive. Iron sources are shown in Table 4. Iron is often supplied to plants by foliar applications in preference to soil applications. Besides the sources shown in the above Table, Fe is also supplied by waste materials, such as sewage sludges, manure, and industrial wastes.

Even though inorganic Fe sources such as finely-ground pyrite and sulfate (Vlek and Lindsay, 1978) have proven effective in some cases, the usual finding is that chelated sources are better. For peanuts,

TABLE 4. Iron fertilizer sources[+]

Source	Formula	Concentration of Fe, g kg^{-1}[‡]
Ferrous sulfate	$FeSO_4 \cdot 7H_2O$	190
Ferric sulfate	$Fe_2(SO_4)_3 \cdot 4H_2O$	230
Ferrous ammonium sulfate	$(NH_4)_2SO_4 \cdot FeSO_4 \cdot 6H_2O$	140
Iron chelate	NaFeEDTA	50-140
Iron chelate	NaFeHEDTA	50-90
Iron chelate	NaFeEDDHA	60
Iron chelate	NaFeDTPA	100
Iron frits	Fritted glass	200-400
Iron lignosulfonate		50-80
Iron polyflavonoid		90-100
Iron methoxyphenylpropane	FeMPP	50

[+]Data from Martens and Westermann (1991).
[‡]Approximate concentration.

FeEDDHA gave the highest yields compared to Fe sulphate, Fe acetate, and several chelate sources (Hartzook, 1982). Soils incubated with Fe chelate sources became lower in soluble Fe at high pH because of fixation and replacement of the chelated Fe by Ca (Boxma, 1981). In that study it was found that at low pH the recovery of soluble Fe decreased in the order: FeDTPA > FeEDTA > FeEDDHA > Fe-HEEDTA. At high pH (7 to 8) the most suitable source was FeEDD-HA, which supports the findings of Hartzook (1982). An FeEDDHA spray increased growth of eucalypts grown on a calcareous soil in Australia whereas FeEDTA spray showed little effectiveness (Stewart et al., 1981).

Besides inorganic and chelate sources for Fe, other methods for supplying Fe, such as using organic wastes and adding acidifying amendments, have been tested experimentally. In a calcareous soil, application of farmyard manure or elemental S significantly increased the availability of Fe (Babaria and Patel, 1980). The same study showed also that saturating the soil with water gave a 3-fold increase in available Fe over a water content at 50% of field capacity. Adding various plant materials to an Fe-deficient soil increased the dry matter yield of sorghum (Parsa, Wallace, and Martin, 1979).

Bioavailability

Soil pH

Iron availability and soil extractability decrease as the soil pH is increased by liming, as shown by Haynes and Swift (1985) for Fe extracted by three methods (Figure 2). Lime-induced chlorosis in a calcareous soil was attributed to the inability of the plants to extract Fe from the soil as opposed to Fe inactivation within the roots (White and Robson, 1989). Since Fe availability is extremely pH sensitive, one of the management techniques to increase Fe supply to plants is to acidify the soil. Additions of elemental S for Bermudagrass reduced soil pH and increased the availability of Fe to the turf (Remon et al., 1977). For a calcareous soil the uptake of Fe and yields of barley were

FIGURE 2. DTPA-extractable Fe as influenced by Fe rate and soil pH. Average 95% confidence interval shown. (Haynes and Swift, 1985, with kind permission from Kluwer Academic Publishers, Dordrecht, The Netherlands).

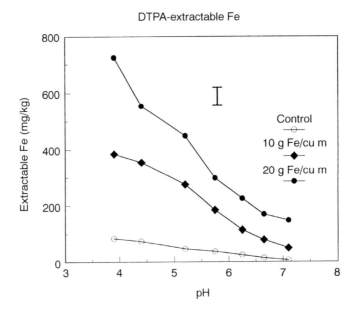

increased by applying 50 mg elemental S kg^{-1} (Al-Ani, Abd-Egawad, and Naji, 1977). For peach trees, additions of acid-forming nitrogenous fertilizers, especially $(NH_4)_2SO_4$, increased Fe concentrations in plant tissues (Cummings and Xie, 1995).

Other Soil Property Effects

Iron availability is influenced to a great extent by the oxidation/reduction condition of the soil, which in turn is affected by such factors as water content, organic matter, and root effects. Lindsay (1991) indicated that the most important influence that organic matter has on the solubilization of Fe is through reduction. Decomposition of organic matter creates microsites, even in well-drained soils, where Fe^{2+} concentration increases above the Fe^{3+} hydrolysis species. Waterlogging and application of organic matter have been shown to increase the water soluble and exchangeable forms of Fe, and the amounts of Fe in these forms increased much more for continuously waterlogged soils than for those alternately waterlogged or saturated (Mandal and Mitra, 1982).

Iron availability has been related to bicarbonate content of soils. Soil pH and bicarbonate had important effects on the availability of Fe and its uptake by peach trees (Koseoglu, 1995). Loeppert (1986) indicated that one of the major factors in enhancing Fe stress is soil solution bicarbonate, which is controlled by the equilibrium relationships with soil carbonates. In soils, the solution phase is as much as 20 times oversaturated with bicarbonate with respect to $CaCO_3$ solubility than in water exposed to the atmosphere (Bloom and Inskeep, 1986). High partial pressures of carbon dioxide leading to higher bicarbonate content in the soil solution are more prevalent under conditions of poor drainage and soil compaction (Loeppert, 1986).

Other soil properties which can influence Fe availability include the ionic strength of the soil solution, the amorphous Fe oxide content, and soil temperature. Morris, Loeppert, and Moore (1990) found that the amorphous Fe oxide content of the soil is an important factor influencing the uptake of Fe by soybeans. They also reported that the Fe deficiency symptoms in soybeans were positively related to the soil solution ionic strength and the reactivity of the soil carbonate phase. High availability of Fe in flooded soils can lead to deficiency of P, K, Ca and Mg as a result of the roots being coated with Fe oxides reducing the root's capacity to absorb nutrients. Inskeep and

Bloom (1986) showed that soil temperatures of 12 and 26°C produced more Fe chlorosis in soybeans at a given soil moisture content than 16 to 19°C.

BORON

Sources

Boron is usually added to soils as a soluble source, such as borax or Solubor® (partially dehydrated borax). Less soluble sources are colemanite and frits, which can be used for sandy, low-organic matter soils where leaching is a problem (Martens and Westermann, 1991). Table 5 shows the commonly used B sources. Waste materials such as manures, sewage sludges, fly ash and effluents can be good sources of B. In some cases, fly ash (El-Mogazi, Lisk, and Weinstein, 1988) and furnace ash (Smidt and Whitton, 1975) can even cause B toxicities. Wood ash has been used as a satisfactory B source (Etiegni et al., 1991).

The forms of B in soil are important in determining plant availability and susceptibility to leaching. Singh and Randhawa (1977) found that 4 to 21% of the total B in saline-alkali soils was water soluble and

TABLE 5. Boron fertilizer sources[†]

Source		Formula	Concentration of B, g kg^{-1}[‡]
Boric acid		H_3BO_3	170
Boron frits		Fitted glass	20-60
Borax		$Na_2B_4O_7 \cdot 10H_2O$	110
Colemanite		$CaB_6O_{11} \cdot 5H_2O$	100
Sodium pentaborate		$Na_2B_{10}O_{16} \cdot 10H_2O$	180
Sodium tetraborate:	Borate-45	$Na_2B_4O_7 \cdot 5H_2O$	140
	Borate-65	$Na_2B_4O_7$	200
Solubor®		$Na_2B_4O_7 \cdot 5H_2O$	200
		$+ Na_2B_{10}O_{16} \cdot 10H_2O$	

[†]Data from Martens and Westermann (1991).
[‡]Approximate concentration.

nearly 25% was leachable. Boron uptake by wheat and bell pepper has been related directly to the B in soil solution rather than to B adsorbed by soil surfaces (Keren, Bingham, and Rhodes, 1985). The appropriate rate of application of B to soybeans for a B-deficient, acid soil was 1 mg kg^{-1} (Datta et al., 1994). Boron was applied to the foliage of cotton and soybeans at a rate of 0.22 kg ha^{-1} using five sources including boric acid and Solubor® (Guertal et al., 1996). There were no differences in tissue B among the sources. The increase in B concentration in cotton tissue from the low to the high concentration was between 8 to 11 mg kg^{-1} and the same increase for soybeans was 16 to 22 mg kg^{-1}.

Bioavailability

Soil pH

Soil B availability to plants decreases with an increase in soil pH, usually brought about by liming. Peterson and Newman (1976) found that B uptake by tall fescue was 2.5 times lower at pH 7.4 than at more acidic pH levels studied. Liming was found to induce B deficiency in peas (Gupta and MacLeod, 1981), and maize and tomatoes (Ambak, Bakar, and Tadano, 1991). The decrease in B in barley and peas caused by increasing lime rates found by Gupta and MacLeod (1981) is shown in Figure 3. Extractable soil B also decreases with increases in soil pH (Khan, Ryan, and Berger, 1979; Kukier, Sumner, and Miller, 1994). The extent of B released from fly ash was influenced mainly by soil pH, with hot-water extractable B being a linear function of soil pH (Kukier, Sumner, and Miller, 1994). In 162 soil samples in Lebanon, only pH was significantly and negatively related to extractable soil B among various soil properties measured (Khan, Ryan, and Berger, 1979). Adding Ca and Mg to soils lowered B uptake by plants, but the decrease was related to soil pH rather than availability of Ca or Mg (Gupta and MacLeod, 1977).

Other Soil Property Effects

Soil properties such as organic matter and clay contents and water potential can influence the availability of B. The organic matter content is positively related to B availability (Gupta, 1978; Lombin,

FIGURE 3. Effect of lime rates on B in barley and pea plant tissue. Values within a crop with different letters are not different at the 5% level according to an LSD. Adapted from Gupta and MacLeod (1981).

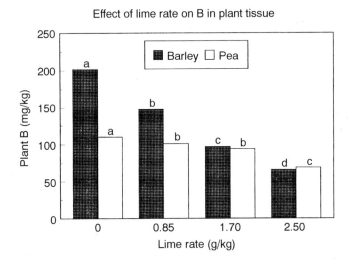

1985). Low B availability to plants was attributed to low organic matter in coarse-textured soils of Nigeria (Lombin, 1985). Tapan, Mondal, and Biswapati (1993) found that with the application of lime and organic matter the B availability increased, probably due to the production of soluble organic compounds that formed complexes with soil B. The amounts of clay in soil are also positively correlated with B availability (Gupta, 1978). Dionne and Pesant (1978) showed that B availability to alfalfa and birdsfoot trefoil was greatly decreased by a combination of repeated drought cycles and by either high (pH 7.4) or low soil pH (< pH 5.0).

MOLYBDENUM

Sources

The main crops that need Mo fertilization are legumes, where Mo plays an important role in symbiotic N fixation and in the reduction of nitrate. The fertilizer sources of Mo are shown in Table 6. Usually, Mo is added as Na molybdate. Rates of Mo application range from 0.01 to

TABLE 6. Molybdenum fertilizer sources[+]

Source	Formula	Concentration of Mo, g kg^{-1}[‡]
Ammonium molybdate	$(NH_4)_6Mo_7O_{24}\cdot4H_2O$	540
Molybdenum trioxide	MoO_3	660
Molybdenum frits	Fritted glass	20-30
Molybdic acid	$H_2MoO_4\cdot H_2O$	530
Sodium molybdate	$Na_2MoO_4\cdot2H_2O$	390

[+]Data from Martens and Westermann (1991).
[‡]Approximate concentration.

0.5 kg ha^{-1}, which is very low compared with the previous microelements discussed (Martens and Westermann, 1991). Yield of alfalfa was increased either by adding Mo or by liming because an increase in soil pH increases Mo availability (Rechcigl, 1987). In some instances liming will suffice to correct Mo deficiency. Jones, Williams, and Ruckman (1977) found responses of subterranean clover to Mo as $(NH_4)_6Mo_7O_{24}\cdot4H_2O$ in 14 of 23 field sites on serpentine soils where Ca, P, and S deficiencies had been remedied first. Fertilizing soybeans with Mo increased yields of nodulated varieties, but non-nodulating varieties did not respond (Hashimoto and Yamasaki, 1976). The latter study showed that the increased yield obtained by the Mo application was due to enhancing the N_2 fixing activity of the root nodules of soybeans.

Bioavailability

Soil pH

Unlike the first five micronutrients discussed, plant availability of Mo increases as soil pH increases. For many crops, liming and/or Mo applications will correct Mo deficiencies. Liming to increase soil pH from 4.3 to 6.9 increased the Mo concentration and yield in red clover, onion, cauliflower, lettuce, and carrot (Gupta, Chipman, and MacKay, 1978). Both lime additions and Mo treatments were required to give maximum corn yields on a soil of pH 4.3 (Tanner and Grant, 1977). For soybeans, alfalfa, and various clovers, dry matter was increased by increasing soil pH on a Mo-deficient soil; response to Mo was at pH 6 and absent at pH 7 (Mortvedt, 1981). Adsorption of Mo by soil colloids is also linked to the soil pH. Tanner (1978) found a significant correlation for Mo adsorption by clays and clay loams with soil pH and exchangeable acidity. The adsorption of Mo by eight Coastal Plain and Piedmont soils was negatively correlated with pH and P levels (Karimian and Cox, 1978).

Other Soil Property Effects

Molybdenum levels in 50 savanna soils of Nigeria were low, which was attributed to low organic matter and the coarse texture of the soils (Lombin, 1985). For coarse-textured Canadian soils, extractable Mo content was positively correlated with clay content (MacLean and Langille, 1983). Even tillage can affect Mo availability (Thomas, 1986), with Mo being less available under no-tillage systems than under conventional tillage, probably due to more acidic conditions in the former. Soybean responses and plant Mo concentrations could be predicted from the active Fe ratio (amorphous Fe/free Fe) and soil pH (Karimian and Cox, 1979). Also, the adsorption of Mo was positively related to the Fe oxide and organic matter contents of the soils studied (Karimian and Cox, 1978) (Figure 4). Interactions of Mo with P, N, and mycorrhiza have been found. Where Mo was applied to Burley tobacco, P application increased Mo plant tissue concentration. Where no Mo was applied, P did not increase plant Mo concentration (Schwamberger and Sims, 1991). Application of ammonium sulfate or calcium ammonium nitrate plus gypsum decreased the Mo content of grasses, while increasing the P rate reduced the Mo levels in the herbage (Feely, 1990). For leucaena grown in an oxisol, mycorrhizal colonization was significantly increased with addition of 4.4 kg Mo ha^{-1} (Aziz and Habte, 1988).

CONCLUSIONS

The micronutrient fertilizer sources that continue to be the most used for Zn, Cu, and Mn are sulfates because of their solubility and wide-spread availability to growers. Chelated Fe is used to prevent Fe from becoming quickly unavailable due to oxidation reactions. For B, the usual source is borax, and for Mo it is sodium molybdate. Soil pH

FIGURE 4. Adsorption isotherms for Mo for five soils with different Fe oxide contents (Karimian and Cox, 1978, with kind permission from ASA-CSSA-SSSA, Madison, Wisconsin, USA).

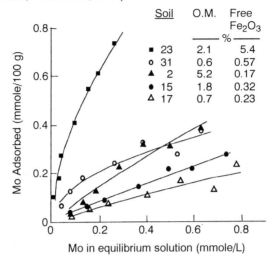

Soil	O.M.	Free Fe$_2$O$_3$
		%
■ 23	2.1	5.4
o 31	0.6	0.57
▲ 2	5.2	0.17
● 15	1.8	0.32
△ 17	0.7	0.23

is certainly the one soil property that most influences micronutrient availability, and for all but Mo, the higher the soil pH, the lower is the plant availability of micronutrients. For Mo, liming can actually prevent deficiencies, since the availability increases with pH. Other soil properties that can be important are organic matter for Cu and oxidation/reduction conditions for Fe and, especially, for Mn.

Future research directions include testing newer sources that are better able to keep the micronutrients in available forms. Also, an important future research direction that is already under way is to test the effects of high loading rates with wastes on short- and long-term availability of micronutrients to plants, and effects on leaching and loss to surface runoff. The additions of soluble organics with organic wastes, and especially manures, can supply micronutrients in forms that are leachable. Thus, the interactions of micronutrient metals with organic materials will continue to be an important research topic. Research is still needed on inorganic micronutrient sources and waste-borne micronutrients to ensure the most economic yields without endangering soil or air quality.

REFERENCES

Adams, J.F., F. Adams, and J.W. Odom. (1982). Interaction of phosphorus rates and soil pH on soybean yield and soil solution composition of two phosphorus-sufficient Ultisols. *Soil Science Society of America Journal* 46: 323-328.

Ahangar, A.G., N. Karimian, A. Abtahi, M.T. Assad, and Y. Emam. (1995). Growth and manganese uptake by soybean in highly calcareous soils as affected by native and applied manganese and predicted by nine different extractants. *Communications in Soil Science and Plant Analysis* 26: 1441-1454.

Al-Ani, F., M. Abd-Egawad, and T. Naji. (1977). Iron and phosphorus availability in soil, and barley yield as influenced by sulfur application. *Iraqi Journal of Agricultural Science* 12: 124-136.

Alley, M.M. (1975). Manganese deficiency in Virginia soils and correction of manganese deficiency in soybeans (*Glycine max* L.). *Dissertation Abstract International* 36: 523.

Alley, M.M., C.I. Rich, G.W. Hawkins, and D.C. Martens. (1978). Corrections of Mn deficiency of soybeans. *Agronomy Journal* 70: 35-38.

Alvarez, J.M., A. Obrador, and M.I. Rico. (1996). Effects of chelated zinc, soluble and coated fertilizers, on soil zinc status and zinc nutrition of maize. *Communications in Soil Science and Plant Analysis* 27: 7-19.

Ambak, K., Z.A. Bakar, and T. Tadano. (1991). Effect of liming and micronutrient application on the growth and occurrence of sterility in maize and tomato plants in a Malaysian deep peat soil. *Soil Science and Plant Nutrition* 37: 689-698.

Aziz, T. and M. Habte. (1988). Role of molybdenum in establishing mycorrhizal leucaena in an eroded oxisol. *Leucaena Research Reports* 9: 109-110.

Babaria, C.J. and C.L. Patel. (1980). Effect of application of iron, farmyard manure and sulphur on the availability of iron in medium black calcareous soil at different moisture regimes. *Journal of the Indian Society of Soil Science* 28: 302-306.

Baker, D.E. and J.P. Sneft. (1995). Copper. In *Heavy Metals in Soils*, ed. B.J. Alloway. New York, USA: Blackie Academic and Professional, pp. 179-205.

Beverly, R.B. (1987). Celery response to foliar nutritional sprays and acidification of a Histosol. *HortScience* 22: 1271-1273.

Blamey, F.P.C. and J. Chapman. (1979). Soil amelioration effects on boron uptake by sunflowers. *Communications in Soil Science and Plant Analysis* 10: 1057-1066.

Bloom, P.R. and W.P. Inskeep. (1986). Factors affecting bicarbonate chemistry and iron chlorosis in soils. *Journal of Plant Nutrition* 9: 215-228.

Boswell, F.C., M.B. Parker, and T.P. Gaines. (1989). Soil zinc and pH effects on zinc concentrations of corn plants. *Communications in Soil Science and Plant Analysis* 20: 1575-1600.

Boxma, R. (1981). Effect of pH on the behavior of various iron chelates in sphagnum (moss) peat. *Communications in Soil Science and Plant Analysis* 12: 755-763.

Boyle, F.W. Jr. and W.L. Lindsay. (1986). Manganese phosphate equilibrium relationships in soils. *Soil Science Society of America Journal* 50: 588-593.

Brennan, R.F. (1994). The residual effectiveness of previously applied copper fertilizer for grain yield of wheat grown on soils of south-west Australia. *Fertilizer Research* 39: 11-18.

Brennan, R.F. and J.W. Gartrell. (1990). Reaction of zinc with soil affecting its availability to subterranean clover. I. The relationship between critical concentrations of extractable zinc and properties of Australian soils responsive to applied zinc. *Australian Journal of Soil Research* 28: 293-302.

Cox, F.R. (1990). A note on the effect of soil reaction and zinc concentration on peanut tissue zinc. *Peanut Science* 17: 15-17.

Cox, F.R. (1992). Residual value of copper fertilization. *Communications in Soil Science and Plant Analysis* 23: 101-112.

Cummings, G.A. and H.S. Xie. (1995). Effect of soil pH and nitrogen source on the nutrient status in peach: II. Micronutrients. *Journal of Plant Nutrition* 18: 553-562.

Datta, S.P., A. Kumar, R.P. Singh, K.P. Singh, and A.K. Sarkar. (1994). Critical limit of available boron for soybean in acid sedentary soils of Chotanagpur region. *Journal of the Indian Society of Soil Science* 42: 93-96.

Davis-Carter, J.G., M.B. Parker, and T.P. Gaines. (1991). Interaction of soil zinc, calcium, and pH with zinc toxicity in peanuts. In *Proceedings of the Second International Symposium on Plant-Soil Interactions at Low pH*, eds. R.J. Wright, V.C. Baligar, and R.P. Murrmann. Dordrecht, The Netherlands: Kluwer Academic Publishers, pp. 339-347.

Davis-Carter, J.G. and L.M. Shuman. (1993). Influence of texture and pH of kaolinitic soils on zinc fractions and zinc uptake by peanuts. *Soil Science* 155: 376-384.

Dionne, J.L. and A.R. Pesant. (1978). Effects of doses of boron, moisture regimes and soil pH on yields of lucerne and birdsfoot trefoil and on availability of boron. *Canadian Journal of Soil Science* 58: 369-379.

El-Mogazi, D., D.J. Lisk, and H. Weinstein. (1988). A review of physical, chemical, and biological properties of fly ash and effects on agricultural ecosystems. *The Science of the Total Environment* 74: 1-37.

Etiegni, L., R.L. Mahler, A.G. Campbell, and B. Shafii. (1991). Evaluation of wood ash disposal on agricultural land. II. Potential toxic effects on plant growth. *Communications in Soil Science and Plant Analysis* 22: 257-267.

Feely, L. (1990). Agronomic effectiveness of nitrogen, sulphur and phosphorus for reducing molybdenum uptake by herbage grown in peatland. *Irish Journal of Agricultural Research* 2: 129-139.

Gallagher, P.J., L.S. Murphy, and R. Ellis, Jr. (1978). Effects of temperature and soil pH on effectiveness of four zinc fertilizers. *Communications in Soil Science and Plant Analysis* 9: 115-126.

Gambrell, R.P. and W.H. Patrick, Jr. (1989). Cu, Zn, and Cd availability in a sludge-amended soil under controlled pH and redox potential conditions. In *Ecological Studies: Inorganic Contaminants in the Vadose Zone*, eds. Y.B. Bar, N.J. Barrow, and J. Goldshmid. Stuttgart, Germany: Springer-Verlag, pp. 89-106.

Gammon, N. Jr. (1976). Zinc and manganese levels in successive saturated soil solution extracts from a Zn or Mn-treated Leon fine sand. *Proceedings of Soil and Crop Science Society of Florida* 35: 98-101.

Gough, L.P., J.M. McNeil, and R.C. Severson. (1980). Predicting native plant copper, iron, manganese, and zinc levels using DTPA and EDTA soil extractants, Northern Great Plains. *Soil Science Society of America Journal* 44: 1030-1036.

Guertal, E.A., A.O. Abaye, B.M. Lippert, G.S. Miner, and G.J. Gascho. (1996). Sources of boron for foliar fertilization of cotton and soybean. *Communications in Soil Science and Plant Analysis* 27: 2815-2828.

Gupta, S.K. (1978). Effect of soil properties on the extractable boron contents. *Schweizerische Landwirtschaftliche Forschung* 17: 45-50.

Gupta, U.C., E.W. Chipman, and D.C. MacKay. (1978). Effects of molybdenum and lime on the yield and molybdenum concentration of crops grown on acid sphagnum peat soil. *Canadian Journal of Plant Science* 58: 983-992.

Gupta, U.C. and J.A. MacLeod. (1977). Influence of calcium and magnesium sources on boron uptake and yield of alfalfa and rutabagas as related to soil pH. *Soil Science* 124: 279-284.

Gupta, U.C. and J.A. MacLeod. (1981). Plant and soil boron as influenced by soil pH and calcium sources on podzol soils. *Soil Science* 131: 20-25.

Hartzook, A. (1982). The problem of iron deficiency in peanuts (*Arachis hypogaea* L.) on basic and calcareous soils in Israel. *Journal of Plant Nutrition* 5: 923-926.

Hashimoto, K. and S. Yamasaki. (1976). Effects of molybdenum application on the yield, nitrogen nutrition and nodule development of soybeans. *Soil Science and Plant Nutrition* 22: 435-443.

Haynes, R.J. and R.S. Swift. (1985). Effects of liming on the extractability of Fe, Mn, Zn and Cu from a peat medium and the growth and micronutrient uptake of highbush blueberry plants. *Plant and Soil* 84: 213-223.

Hue, N.V. (1988). A possible mechanism for manganese phytotoxicity in Hawaii soils amended with a low-manganese sewage sludge. *Journal of Environmental Quality* 17: 473-479.

Indulkar, B.S. and G.U. Malewar. (1994). Response of sorghum (*Sorghum bicolor*) to different zinc sources and their residual effect on succeeding wheat (*Triticum aestivum*). *Indian Journal of Agronomy* 39: 368-372.

Inskeep, W.P. and P.R. Bloom. (1986). Effects of soil moisture on soil pCO_2, soil solution bicarbonate and iron chlorosis in soybeans. *Soil Science Society of America Journal* 50: 945-952.

Iwasaki, K., G. Yoshikawa, and K. Sakurai. (1993). Fractionation of zinc in greenhouse soils. *Soil Science and Plant Nutrition* 39: 507-515.

Jahiruddin, M., B.J. Chambers, M.S. Cresser, and N.T. Livesey. (1992). Effects of soil properties on the extraction of zinc. *Geoderma* 52: 199-208.

Jones, M.B., W.A. Williams, and J.E. Ruckman. (1977). Fertilization of *Trifolium subterraneum* L. growing on serpentine soils. *Soil Science Society of America Journal* 41: 87-89.

Jugsujinda, A. and W.H. Patrick, Jr. (1977). Growth and nutrient uptake by rice in a flooded soil under controlled aerobic-anaerobic and pH conditions. *Agronomy Journal* 69: 705-710.

Jurkowska, H. and A. Rogoz. (1977). The influence of liming on copper concentration in plants in relation to the dose and kind of copper fertilizer. *Polish Journal of Soil Science* 10: 149-156.

Karimian, N. and F.R. Cox. (1978). Absorption and extractability of molybdenum in relation to some chemical properties of soil. *Soil Science Society of America Journal* 42: 757-761.

Karimian, N. and F.R. Cox. (1979). Molybdenum availability as predicted from selected soil chemical properties. *Agronomy Journal* 71: 63-65.

Karimian, N. and J. Yasrebi. (1995). Prediction of residual effects of zinc sulfate on growth and zinc uptake of corn plants using three zinc soil tests. *Communications in Soil Science and Plant Analysis* 26: 277-287.

Keren, R., F.T. Bingham, and J.D. Rhoades. (1985). Plant uptake of boron as affected by boron distribution between liquid and solid phases in soil. *Soil Science Society of America Journal* 49: 297-302.

Khan, Z.D., J. Ryan, and K.C. Berger. (1979). Available boron in calcareous soils of Lebanon. *Agronomy Journal* 71: 688-690.

Kingery, W.L., C.W. Wood, D.P. Delaney, J.C. Williams, G.L. Mullins, and E. Van Santen. (1993). Implications of long-term land application of poultry litter on tall fescue pastures. *Journal of Production Agriculture* 6: 390-395.

Koseoglu, A.T. (1995). Investigation of relationships between iron status of peach leaves and soil properties. *Journal of Plant Nutrition* 18: 1845-1859.

Kukier, U., M.E. Sumner, and W.P. Miller. (1994). Boron release from fly ash and its uptake by corn. *Journal of Environmental Quality* 23: 596-603.

Levesque, M.P. and S.P. Mathur. (1986). Soil tests for copper, iron, manganese, and zinc in Histosols: 1. The influence of soil properties, iron, manganese, and zinc on the level and distribution of copper. *Soil Science* 142: 153-163.

Li, Z. and L.M. Shuman. (1996). Redistribution of forms of zinc, cadmium, and nickel in soils treated with EDTA. *The Science of the Total Environment* 191: 95-107.

Liang, J., R.E. Karamanos, and J.W.B. Stewart. (1991). Plant availability of Zn fractions in Saskatchewan soils. *Canadian Journal of Soil Science* 71: 507-517.

Lindsay, W.L. (1991). Iron oxide solubilization by organic matter and its effect on iron availability. *Plant and Soil* 130: 27-34.

Lins, I.D.G. and F.R. Cox. (1988). Effect of soil pH and clay content on the zinc soil test interpretation for corn. *Soil Science Society of America Journal* 52: 1681-1685.

Loeppert, R.H. (1986). Reactions of iron and carbonates in calcareous soils. *Journal of Plant Nutrition* 9: 195-214.

Lombin, G. (1985). Evaluation of the micronutrient fertility of Nigeria's semi-arid savannah soils: boron and molybdenum. *Soil Science and Plant Nutrition* 31: 13-25.

Ma, Y.B. and N.C. Uren. (1996). The effects of cropping corn on the extractability of zinc added to a calcareous soil. *Plant and Soil* 181: 221-226.

MacLean, K.S. and W.M. Langille. (1983). The extractable trace element content of coarse textured Annapolis Valley soils and the influence of pH, organic matter and clay content. *Communications in Soil Science and Plant Analysis* 14: 463-469.

McBride, M.B. (1994). *Environmental Chemistry of Soils*. New York, USA: Oxford University Press.

McLaren, R.G. and G.S.P. Ritchie. (1993). The long-term fate of copper fertilizer applied to a lateritic sandy soil in Western Australia. *Australian Journal of Soil Research* 3: 39-50.

Maji, B., S. Chatterji, and B.K. Bandyopadhyay. (1993). Available iron, manganese, zinc and copper in coastal soils of Sundarbans, West Bengal in relation to soil characteristics. *Journal of the Indian Society of Soil Science* 41: 468-471.

Mandal, L.N. and R.R. Mitra. (1982). Transformation of iron and manganese in rice soils under different moisture regimes and organic matter applications. *Plant and Soil* 69: 45-56.

Martens, D.C. and D.T. Westermann. (1991). Fertilizer applications for correcting micronutrient deficiencies. In *Micronutrients in Agriculture*, eds. J.J. Mortvedt, F.R. Cox, L.M. Shuman, and R.M. Welch, Madison, Wisconsin, USA: Soil Science Society of America, pp. 549-592.

Mathur, S.P. and M.P. Levesque. (1989). Soil tests for copper, iron, manganese, and zinc in Histosols: 4. Selection on the basis of soil chemical data and uptakes by oats, carrots, onions, and lettuce. *Soil Science* 148: 424-432.

May, G.M. and M.P. Pritts. (1993). Phosphorus, zinc, and boron influence yield components in 'Earliglow' strawberry. *Journal of the American Society of Horticultural Science* 118: 43-49.

Modaihsh, A.S. (1990). Zinc diffusion and extractability as affected by zinc carrier and soil chemical properties. *Fertilizer Research* 25: 85-91.

Moraghan, J.T. (1996). Zinc concentration of navy bean seedlings as affected by rate and placement of three zinc sources. *Journal of Plant Nutrition* 19: 1413-1422.

Morris, D.R., R.H. Loeppert, and T.J. Moore. (1990). Indigenous soil factors influencing iron chlorosis of soybean in calcareous soils. *Soil Science Society of America Journal* 54: 1329-1336.

Mortvedt, J.J. (1981). Nitrogen and molybdenum uptake and dry matter relationships of soybeans and forage legumes in response to applied molybdenum on acid soil. *Journal of Plant Nutrition* 3: 245-256.

Mozaffari, M., A.K. Alva, and E.Q. Chen. (1996). Relation of copper extractable from soil and pH to copper content and growth of two citrus rootstocks. *Soil Science* 161: 786-792.

Neilsen, D., G.H. Neilsen, A.H. Sinclair, and D.J. Linehan. (1992). Soil phosphorus status, pH and the manganese nutrition of wheat. *Plant and Soil* 145: 45-50.

Ohki, K., F.C. Boswell, M.B. Parker, L.M. Shuman, and D.O. Wilson. (1987). Foliar manganese application to soybeans. *Communications in Soil Science and Plant Analysis* 18: 243-253.

Parker, M.B., T.P. Gaines, C.O. Plank, J.B. Jones, Jr., M.E. Walker, F.C. Boswell, and C.C. Dowler. (1988). *The effects of limestone on corn, soybeans, and peanuts grown on a Tifton soil at various intervals for seventeen years*. Research Bulletin, Georgia Agricultural Experiment Stations. No. 373.

Parker, M.B., T.P. Gaines, M.E. Walker, C.O. Plank, and J.G. Davis-Carter. (1990). Soil zinc and pH effects on leaf zinc and the interaction of leaf calcium and zinc on zinc toxicity of peanuts. *Communications in Soil Science and Plant Analysis* 21: 2319-2332.

Parker, M.B. and M.E. Walker. (1986). Soil pH and manganese effects on manganese nutrition of peanut. *Agronomy Journal* 78: 614-620.

Parsa, A.A., A. Wallace, and J.P. Martin. (1979). Enhancement of iron availability by some organic materials. *Journal of Agricultural Science (Cambridge)* 93: 115-120.

Peaslee, D.E. (1980). Effect of extractable zinc, phosphorus, and soil pH on zinc concentrations in leaves of field-grown corn. *Communications in Soil Science and Plant Analysis* 11: 417-425.

Peterson, L.A. and R.C. Newman. (1976). Influence of soil pH on the availability of added boron. *Soil Science Society of America Journal* 40: 280-282.

Petrie, S.E. and T.L. Jackson. (1984). Effects of fertilization on soil solution pH and manganese concentration. *Soil Science Society of America Journal* 48: 315-318.

Petruzzelli, G., G. Guidi, and L. Lubrano. (1978). Organic matter as an influencing factor on copper and cadmium adsorption by soils. *Water, Air, and Soil Pollution* 9: 263-269.

Raghbir, S., U.C. Shukla, and R. Singh. (1976). Studies of ^{65}Zn movement in soil columns under laboratory conditions. *Geoderma* 15: 313-321.

Reed, S.T., M.G. Allen, D.C. Martens, and J.R. McKenna. (1993). Copper fractions extracted by Mehlich-3 from soils amended with either $CuSO_4$ or copper rich pig manure. *Communications in Soil Science and Plant Analysis* 24: 827-839.

Rechcigl, J.E. (1987). Alfalfa growth on acid soil as influenced by Al, Ca, pH and Mo. *Dissertation Abstracts International B* 47: 4354B.

Remon, J., T. Goodale, J. Ward, and C.E. Whitcomb. (1977). Effects of sulfur on reducing the pH of an alkaline soil. Research Report, Agricultural Experiment Station, Oklahoma State University, No. 760, pp. 18-20.

Sangwan, B.S. and S. Kuldeep. (1993). Vertical distribution of Zn, Mn, Cu and Fe in the semi-arid soils of Haryana and their relationships with soil properties. *Journal of the Indian Society of Soil Science* 41: 463-467.

Schwamberger, E.C. and J.L. Sims. (1991). Effects of soil pH, nitrogen source, phosphorus, and molybdenum on early growth and mineral nutrition of Burley tobacco. *Communications in Soil Science and Plant Analysis* 22: 641-657.

Sedberry Jr., J.E., F.J. Peterson, F.E. Wilson, D.B. Mengel, P.E. Schilling, and R.H. Brupbacher. (1980). Influence of soil reaction and applications of zinc on yields and zinc contents of rice plants. *Communications in Soil Science and Plant Analysis* 11: 283-295.

Shang, C. and T.E. Bates. (1987). Comparison of zinc soil tests adjusted for soil and fertilizer phosphorus. *Fertilizer Research* 11: 209-220.

Sharma, K.N. and D.L. Deb. (1991). Uptake of zinc by wheat plants in relation to their diffusion coefficients under varying physical and chemical environment. *Journal of Nuclear Agriculture and Biology* 20: 102-107.

Sharma, B.D., P.S. Sidhu, and V.K. Nayyar. (1992). Distribution of micronutrients in arid zone soils of Punjab and their relation with soil properties. *Arid Soil Research and Rehabilitation* 6: 233-242.

Shuman, L.M. (1988). Effect of organic matter on the distribution of manganese, copper, iron, and zinc in soil fractions. *Soil Science* 146: 192-198.

Shuman, L.M., F.C. Boswell, K. Ohki, M.B. Parker, and D.O. Wilson. (1979). Soybean yield, leaf manganese, and soil manganese as affected by sources and rates of manganese and soil pH. *Agronomy Journal* 71: 989-991.

Singh, J. and N.S. Randhawa. (1977). Boron fractionation and mineral composition of saline-alkali soils. *Journal of the Indian Society of Soil Science* 25: 433-435.

Smidt, R.E. and J.S. Whitton. (1975). Note on boron toxicity in a stand of radiata pine in Hawkes Bay. *New Zealand Journal of Science* 18:109-113.

Snyder, G.H., E.O. Burt, and G.J. Gascho. (1979). Correcting pH-induced manganese deficiency in Bermudagrass turf. *Agronomy Journal* 71: 603-608.

Stewart, H.T.L., D.W. Flinn, P.J. Baldwin, and J.M. James. (1981). Diagnosis and correction of iron deficiency in planted eucalypts in north-west Victoria. *Australian Forest Research* 11: 185-190.

Tadesse, W., J.W. Shuford, R.W. Taylor, D.C. Adriano, and K.S. Sajwan. (1991). Comparative availability to wheat of metals from sewage sludge and inorganic salts. *Water, Air, and Soil Pollution* 55: 397-408.

Tagwira, F., M. Phia, and L.M. Mugwira. (1993). Zinc studies in Zimbabwean soils: effect of lime and phosphorus on growth, yield and zinc status of maize. *Communications in Soil Science and Plant Analysis* 24: 717-736.

Tanner, P.D. (1978). Relations of sorption of molybdate and phosphate by clays and clay loams to soil pH and other chemical factors. *Rhodesian Journal of Agricultural Research* 16: 31-41.

Tanner, P.D. and P.M. Grant. (1977). Response of maize (*Zea mays* L.) to lime and molybdenum on acid red and yellow-brown clays and clay loams. *Rhodesian Journal of Agricultural Research* 15: 143-149.

Tapan, A., A.K. Mondal, and M. Biswapati. (1993). Influence of organic matter and lime application on boron availability in soils. *Indian Journal of Agricultural Science* 63: 803-806.

Thomas, G.W. (1986). Mineral nutrition and fertilizer placement. In *No Tillage and Surface-Tillage Agriculture: The Tillage Revolution*, eds. M.A. Sprague and G.B. Triplett. New York, USA: John Wiley & Sons, pp. 93-148.

Van Breemen, N., C.C. Quijano, and S. Le-Ngoc. (1980). Zinc deficiency in wetland rice along a toposequence of hydromorphic soils in the Philippines. I. Soil conditions and hydrology. *Plant and Soil* 57: 203-214.

Van Eysinga, J.P.N.L.R., P.A. Van Dijk, and S.S. De Bes. (1978). Available manganese content of soils in the Netherlands determined by various methods. *Communications in Soil Science and Plant Analysis* 9: 141-151.

Velez-Ramos, A. and L. Standifer. (1977). Effect of organic matter and moisture level on extractable soil manganese. *Journal of Agriculture of the University of Puerto Rico* 61: 115-125.

Vlek, P.L.G. and W.L. Lindsay. (1978). Potential use of finely disintegrated iron pyrite in sodic and iron-deficient soils. *Journal of Environmental Quality* 7: 111-114.

White, M.C. and R.L. Chaney. (1980). Zinc, cadmium, and manganese uptake by soybean from two zinc- and cadmium-amended Coastal Plain soils. *Soil Science Society of America Journal* 44: 308-313.

White, P.F. and A.D. Robson. (1989). Poor soil aeration or excess soil CaCO$_3$ induces Fe deficiency in lupins. *Australian Journal of Agricultural Research* 40: 75-84.

Wood, B.W. and A.W. White, Jr. (1986). Influence of disk cultivation and subsoiling on productivity of a mature pecan orchard. *HortScience* 21: 66-68.

Wu, X. and I. Aasen. (1994). Models for predicting soil zinc availability for barley. *Plant and Soil* 163: 279-285.

SUBMITTED: 02/14/97
ACCEPTED: 07/10/97

Delivering Fertilizers Through Seed Coatings

James M. Scott

SUMMARY. Coating seeds with nutrients may be an effective way of supplying starter fertilizer for establishment of seedlings and early growth. However, a large variability in benefits of providing nutrients with seeds has been reported, highlighting differences in the implementation of the techniques. This chapter will discuss the main principles relating to the imprecisely defined concept of nutrient-amended seeds. *[Article copies available for a fee from The Haworth Document Delivery Service: 1-800-342-9678. E-mail address: getinfo@haworthpressinc.com]*

KEYWORDS. Fertilizer, germination, nutrient, seed coating, seed soaking, seedling growth

INTRODUCTION

There has been a continuing interest in the scientific literature over many decades concerning ways of manipulating seed performance

James M. Scott, Division of Agronomy & Soil Science, School of Rural Science and Natural Resources, University of New England, Armidale NSW 2351, Australia (E-mail: jscott@metz.une.edu.au).

Detailed research investigations by colleagues have contributed greatly to this chapter. Specific acknowledgment is warranted for the efforts of Mr. Derek Cameron, Dr. Julie Ascher, Mr. Brian Rubzen, Assoc. Prof. Graeme Blair, Assoc. Prof. Robin Jessop and Dr. Alan Andrews, as well as the excellent technical assistance of Ms. Anne White, Mr. Greg Chamberlain and Ms. L. Lisle. Financial support from the Wool Research and Development Corporation, the Meat Research Corporation, the Rural Industries Research and Development Corporation, and the University of New England, which supported many of these investigations, is gratefully acknowledged.

[Haworth co-indexing entry note]: "Delivering Fertilizers Through Seed Coatings." Scott, James M. Co-published simultaneously in *Journal of Crop Production* (Food Products Press, an imprint of The Haworth Press, Inc.) Vol. 1, No. 2 (#2), 1998, pp. 197-220; and: *Nutrient Use in Crop Production* (ed: Zdenko Rengel) Food Products Press, an imprint of The Haworth Press, Inc., 1998, pp. 197-220. Single or multiple copies of this article are available for a fee from The Haworth Document Delivery Service [1-800-342-9678, 9:00 a.m. - 5:00 p.m. (EST). E-mail address: getinfo@haworthpressinc.com].

through seed treatments and coatings. The purposes for applying these seed treatments have been broad, including the incorporation of rhizobia, the protection from insects, mites and disease, improved handling characteristics, the stimulation of germination and vigour and the delivery of nutrients. This chapter covers the addition of nutrients to seeds and hence will refer to both soaking and coating seeds with nutrients or fertilizers.

Seed soaking has been defined as a process "by which seeds can be led to absorb nutrients, protectants, growth regulators, etc. by immersing them in appropriate solutions for extended periods." It is thus a term covering a wide range of solutions with process variables that are rarely defined by those reporting results. A definition of seed coating is "a general term for the application of finely ground solids or liquids containing dissolved or suspended solids to form a more or less continuous layer covering the natural seed coat and includes pelleting and many other seed treatments" (Scott, 1989a). Again, this definition is broad, and includes a wide range of processes that seeds have been subjected to. In the case of nutrient coatings, the chemical and physical properties of materials added to seeds are often not described. It is therefore likely that some supposedly similar coatings reported in the literature, and prepared by different researchers, are in fact quite different. This chapter will attempt to highlight the main principles which are understood about these imprecisely defined nutrient-amended seeds.

One of the principal attractions of adding nutrients to seeds is the opportunity this provides for improving the efficiency of fertilizer use and for providing the sown seed with its own "package" of nutrients, thereby perhaps supplying it with a competitive advantage during establishment. Thus, with nutrient coatings, the emphasis has been on "starter" fertilizers for establishment rather than on using seed coatings to deliver broad-scale fertilizer applications. Soaking in nutrient solutions has been used as a means of delivering sufficient trace elements and of stimulating plant growth by a process which not only introduces macronutrients but also may be confounded with "priming" effects (Heydecker and Coolbear, 1977).

The placement of fertilizer in relation to sown seeds is of special importance for immobile nutrients such as phosphorus, potassium and manganese (Barber and Kovar, 1985; Welch et al., 1966) as well as copper, iron and zinc (Mortvedt, 1994). Because these elements move little in the soil, roots need to grow into the vicinity of the fertilizer source.

During establishment, it is crucial that seedlings emerge promptly and then continue to grow rapidly, especially in relation to any weeds establishing at the same time. This is true of crops, pastures and trees during establishment. Small seeded species can respond to a source of external nutrients as early as four days after sowing; hence roots need to reach a source of sufficient nutrients in the soil soon after germination. The nutrient-amended seed provides a vehicle for delivering controlled quantities of fertilizer to aid early access to nutrients by seedlings.

Nutrient amendment aims to increase the supply of nutrients over that contained in natural seeds. Work by Bolland and Baker (1988), in nutrient-deficient sandy soils in Western Australia, has shown that the nutrient content of seeds can be important when that seed is produced in, and then re-sown into, extremely nutrient-deficient soil. However, in most agricultural soils, nutrient deficiencies in the soil lead to lower seed yields rather than nutrient-deficient seed. Seed soaking with nutrients can provide establishing plants with elevated levels of critical nutrients, whilst seed coating provides the opportunity to supply far higher levels of nutrition than can be achieved in natural seeds.

Reviews of seed coating which have included some information on nutrient coatings have been published by Heydecker and Coolbear (1977) and Scott (1989a). Other reviews of seed treatments give scant attention to this area. Seed coating has for a long time been regarded as an "art" or skill; the patent literature attests to this in the many inventions recorded there, many of which have never been commercialised. This may, in part, be due to the claims being made for some inventions not being borne out by subsequent objective evaluations.

Because there are so many interactions between the species, the seed, its environment and nutrient amendment, there is a need to better understand the science behind seed soaking and coating if reliable performance is to be attained for a wide range of species.

BENEFITS OF SEED SOAKING AND COATING

The effects of soaking and coating can be quite different. In cases where soaking has been successful, the major benefit observed appears to be related either to the supply of a trace element or to a biochemical stimulus to growth and development. Successful nutrient coatings, on the other hand, promote growth by supplying an external source of nutrients that is sufficient to support at least early seedling growth.

Applying active materials to seeds offers the opportunity to supply nutrients adjacent to each and every seed in a uniform fashion. This can lead not only to increasing early seedling growth (e.g., Silcock and Smith, 1982) but can also lead to more uniform seedlings compared to drilled applications of fertilizer (Scott and Blair, 1988a, b). In spite of nutrient coatings slowing early emergence, Miller et al. (1971) stated that the advantages of seed nutrient coating may outweigh any disadvantages associated with any delay in emergence. Nevertheless, nutrient-coated seeds have still not been widely adopted.

In addition to the opportunity to supply seedlings with fertilizer, seed coatings have been claimed to have other benefits. These include better handling through machinery (especially for seed with hairy appendages such as some grasses), improved germination (chiefly with relatively inert coatings) (Dowling, Clements, and McWilliam, 1971) and improved "ballistics" when surface sown (Vartha and Clifford, 1973). If seeds are coated with at least one material, then the opportunity is presented by that process for multiple components (e.g., the addition of fungicides and/or insecticides to nutrient coatings) to be added to the seed as long as the materials added are all compatible with each other.

A study of interactions between nutrient-coated phalaris and an annual grass weed was carried out in a de Wit style diallele replacement series experiment, where increasing sowing rates of phalaris were planted with decreasing sowing rates of the weed. The results showed that phosphorus coatings applied to phalaris seeds resulted in the phosphorus being more available to the phalaris than to the annual grass weeds sown in the same pot (Wythes, 1990).

An example of the effects of micronutrient supply was reported in high pH soil which was severely deficient in manganese, where sowing seeds high in manganese content led to increased plant survival, growth and yield (Longnecker, Marcar, and Graham, 1991). The elevated seed manganese content was equally successful, whether the seed was naturally rich in manganese as a result of the nutrition of the mother plant, or if it was due to soaking in manganese sulfate. A further example of benefits from manganese is the reduction of the disease take-all of wheat by the application of manganese sulfate to seeds (Wilhelm, Graham, and Rovira, 1988).

NUTRIENT INTERACTIONS WITH SEEDS

Most of the literature concerning nutrient effects on seeds concerns firstly the effects of those nutrients on germination and stand establishment and secondly on growth enhancement as a result of the nutrients contained in the seed coatings.

Effects on Germination and Emergence

Toxicity

Whilst fertilizer injury was claimed by Rader, White, and Whittaker (1943) to be due to osmotic effects of fertilizers, much of the evidence suggests that they are in fact toxic to seeds (Carter, 1969). Others have suggested that seeds in contact with fertilizer suffer a "physiological drought," but Philip (1958) also found that damage to seeds in contact with fertilizer was due to the entry of toxic quantities of solutes into seeds.

In the case of phosphorus, toxicities occur when uptake results in tissue concentrations greater than 1% by dry weight. High rates of phosphorus uptake can severely affect plant stands (Olsen and Dreier, 1956; Richards, Batos, and Sheppard, 1985). Lower rates may not be lethal, but the slowing of emergence by seed coatings is commonly reported.

Some soluble ions such as nitrate, chloride and sodium can also be toxic to seeds. Injury due to toxicity appears to be greatest when soil moisture is low and hence solute concentration is high (Guttay, 1957; Carter, 1969).

In the case of injury from nitrogen, it appears that free ammonia is the main toxic agent. Ensminger, Hood, and Willis (1965) showed that diammonium phosphate (DAP) was toxic due to the release of ammonia, especially at high soil pH. Subsequent treatment with magnesium sulfate overcame some of the negative effects of soaking in ammonia solutions. Because of this, it was suggested that DAP may act by inactivating enzyme systems due to precipitating magnesium.

Fertiliser toxicity during germination is difficult to observe as seeds generally die before emerging from the soil. Once germinated, injury to seedlings usually occurs to seminal roots. Fertilizer injury appears to be greatest when soil temperatures are high as there is less opportunity for soils to sorb fertilizer prior to contact with plant roots.

A positive correlation was found between pH of the coating and germination in glasshouse trials by Scott (1975) and Scott and Blair (1988a) but, in contrast, there was no effect found in the field by Scott (1975) nor in another pot trial by Scott et al. (1987). Also, Rebafka, Bationo, and Marschner (1993) found no correlation between pH and osmolality of the coatings and final emergence of seedlings. Thus is appears that, if there is any relationship with pH of the coating, then it is highly dependent on fertilizer type and field conditions.

Scott et al. (1987) showed that the emergence of wheat coated with urea was lower at low soil moisture contents. However, with less soluble coatings, Scott, Mitchell, and Blair (1985) found emergence was lowest in wet soils; in this case, it was thought to be due to poor aeration rather than toxicity.

Injury to crop seeds coated with phosphorus fertilizer also tends to be affected by soil texture. For example, damage from monocalcium phosphate was found to be greatest on coarse-textured soil (Ascher, Scott, and Jessop, 1990).

In general, the effectiveness of localised placement of fertilizer depends to a large extent on the sorption capacity of the soil and on the amount of soil with which the fertilizer is mixed. Barber and colleagues (e.g., Anghinoni and Barber, 1980; Barber and Kovar, 1985) have investigated this in great detail, including the modelling of such effects.

Cruciferous species tend to be highly susceptible to fertilizer injury (Carter, 1969). Hayward and Scott (1993) reported that soluble fertilizers drilled with the seed generally had a deleterious effect on the establishment of turnips (*Brassica* spp.). Legumes and graminaceous species tend to be more tolerant than cruciferous species (Ascher, Scott, and Jessop, 1987). Legumes which germinate rapidly, such as lucerne, have been found to be damaged even by low rates of soluble phosphorus coatings (Scott and Blair, 1988a, b).

Guttay (1957) pointed out that fertilizer injury tends to be lower in species that possess the lemma and palea surrounding the seed compared to naked seeds. In fact, when they removed the outer seed parts of buffel grass, Silcock and Smith (1982) found that the caryopses were more susceptible to all soluble phosphorus sources. These authors postulated that the outer seed parts of buffel grass act as a semi-permeable membrane resulting in tolerance of fertilizers coated on the seeds. Garrote, Scott, and Blair (1987) confirmed this phenomenon with five grasses. Further work by Scott and O'Donnell (1987) and Garrote et al. (1990a) showed that it is most likely an effect of slowing emergence long enough for reactions between the soil and fertilizer to take place, thus lowering solute concentrations in the vicinity of the seed.

In studies which have compared different fertilizer elements, nitrogen sources placed close to seeds are generally regarded as most damaging, with potassium ranked second. Phosphorus is considered less damaging (Olson and Dreier, 1956; Carter, 1969; Hoeft, Walsh, and Liegel, 1975), whilst sulfur is considered to be relatively safe to germinating seeds.

MODELS OF COATED SEED AND GERMINATION

Models of the behaviour of coated seed in the literature are rare. The most detailed models concern coatings containing relatively little nutrient and relate mostly to particle size adjacent to the seed coat (Sharples and Gentry, 1980). As pointed out in Scott (1989b), few attempts have been made to extend models of seed germination and establishment to include the influence of seed coatings. As shown in Figure 1, a seed coating can be viewed as a series of concentric layers, each with its own properties in much the same way as Collis-George (1987) described the seed as having three components. The area which requires more in-depth understanding is the interface between the testa of the seed and the inner coating layer; models may be able to assist in identifying the crucial missing parts of our understanding.

PERFORMANCE OF NUTRIENT-ENHANCED SEEDS

Soaking Seeds in Nutrient Solutions

Soaking seeds in nutrient solutions is capable of supplying limited quantities of nutrients. However, with micronutrients, the amounts

FIGURE 1. Diagram of an idealised coated seed indicating the key properties of each seed component and coating shell layer.

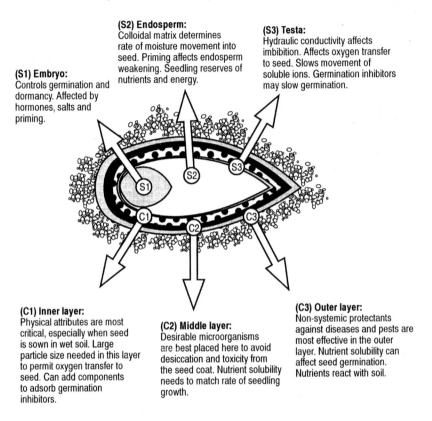

(S2) Endosperm:
Colloidal matrix determines rate of moisture movement into seed. Priming affects endosperm weakening. Seedling reserves of nutrients and energy.

(S3) Testa:
Hydraulic conductivity affects imbibition. Affects oxygen transfer to seed. Slows movement of soluble ions. Germination inhibitors may slow germination.

(S1) Embryo:
Controls germination and dormancy. Affected by hormones, salts and priming.

(C1) Inner layer:
Physical attributes are most critical, especially when seed is sown in wet soil. Large particle size needed in this layer to permit oxygen transfer to seed. Can add components to adsorb germination inhibitors.

(C2) Middle layer:
Desirable microorganisms are best placed here to avoid desiccation and toxicity from the seed coat. Nutrient solubility needs to match rate of seedling growth.

(C3) Outer layer:
Non-systemic protectants against diseases and pests are most effective in the outer layer. Nutrient solubility can affect seed germination. Nutrients react with soil.

imbibed by seeds are sufficient to increase growth and survival and even increase final yield. With macronutrients, soaking supplies relatively little of the plant's requirement (compared to coating) and yet some dramatic results from these soaking treatments have been reported.

For example, under Indian conditions, soaking in 1% KH_2PO_4 produced a significant wheat yield increase (Paul, Choudhury, and Dey, 1993). Similarly, maize seed germination, vigour and seedling growth rate were increased by soaking for 8 h in a KH_2PO_4 solution (Kurdikeri, Aswathariah, and Prasad, 1993). Nitrogen and magnesium also have been shown to affect seedling growth as soaking wheat seeds in magnesium nitrate (5 mM) resulted in increased N content and shoot and root weight of seedlings (Bose and Mishra, 1992).

An experiment with manganese coatings and soaking comparisons was reported by McEvoy et al. (1988) who found that manganese sulfate applied by either method could dramatically increase plant survival and yield in severely manganese-deficient sandy, calcareous soils in South Australia.

The mechanism of how soaking affects plant growth and yield is unclear, but it may well be linked to hormonal changes in seeds such as those brought about by deliberately soaking seeds in growth regulators. Chhipa and Lal (1993) soaked wheat seeds in different growth regulators, resulting in different ionic ratios of N:P, K:P and K:Ca in the grain which were related to an increase in the tolerance of wheat to salinity, apparently enabling the seed to minimize sodium uptake.

Relatively few experiments have been reported on comparisons between the effects of soaking and coating with nutrients. Because of the possible confounding of priming with the effects of small amounts of nutrients added, it is important that experiments with coatings include relevant controls such as seeds soaked for the same length of time in distilled water. Similarly, experiments investigating nutrient-coated seeds should also include a control with seeds coated with an inert material to enable the separation of the physical and chemical effects of the coating.

Coating Seeds with Nutrients

Coating with nutrients can add agronomically significant quantities of fertilizer to establishing crops (e.g., up to 10 kg per ha of some elements). Some aspects of the process and formulation of nutrient coated seeds will be outlined below.

Coating Process

The literature relating to coating processes has been reviewed by Scott (1989a). Some of the most detailed descriptions of coating seeds are contained in the patent literature (e.g., Funsten and Burgesser, 1951). In brief, coating with nutrients usually entails the sequential coating of seeds with layers of adhesive followed by finely ground nutrients until, after sieving, the desired increase in size is achieved.

This is usually followed by careful drying. Most coating is carried out in batches which gives good control over the ratio of seeds to coating materials. Few continuous flow seed coating processes have been used commercially.

Adhesives Used. Commonly, the adhesives used in seed coatings tend to be biologically safe but weak adhesives, such as methyl cellulose. Others, such as polyvinyl alcohol, are somewhat more expensive but have been shown to produce greater pellet integrity following drying. More details are given in Scott (1989a), and a comparison of coating quality achieved with a number of adhesives is given by Scott, Blair, and Andrews (1997).

Pan Coating. Scott, Blair, and Andrews (1997) have published details of some objective studies of the most common method of coating seeds–that of pan coating. The paper highlights the critical importance of the degree of atomisation of the adhesive as well as the importance of selecting an appropriate adhesive. Figure 2 shows the layout of a typical coating, sieving and drying equipment used in the batch processing of coated seeds.

Fluid Bed Coating. Liu and Litster (1993a, b) have conducted fundamental studies of the dynamics of coating seeds with nutrients using fluid bed technology. This process maintains a "bed" of seeds agitating on rapidly moving air whilst adhesive and nutrient slurry are injected near the base of the bed. Coating and drying are carried out simultaneously. Using this process, they were able to produce single coated seeds with high coating integrity and with equivalent agronomic performance to seeds coated by a more conventional pan coating technique. Fluid-bed coating lends itself to continuous flow coating which has the potential to lower costs of coating and produce greater uniformity of coated seeds.

Nutrient Formulations

Most coating formulations used by commercial seed-coating firms are closely guarded secrets and thus are not able to be discussed here. In general terms, it is common for coatings to contain lime as well as fertilizer materials and perhaps fillers such as bentonite. When lime is included, it usually results in at least some reversion of the soluble components to insoluble components, such as with reverted superphosphate (in which phosphorus is present as dicalcium phosphate).

FIGURE 2. Diagram of equipment used for the pan coating of seeds in a commercial seed coating facility.

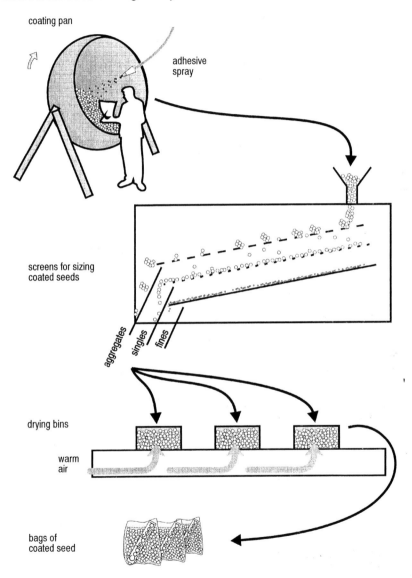

Some details from the literature on the types of fertilizer applied in coatings are given below.

Phosphorus. Phosphorus coatings have been experimented with over many years. One of the early attempts was by Guttay et al. (1957) who examined phosphorus coatings on corn seed. The final stand of lettuce was enhanced by the addition of phosphorus (P) in tablets (Sharples and Gentry, 1980). Scott et al. (1987) found that monocalcium phosphate (MCP) coatings were more damaging to the emergence of wheat than oats. Scott and Blair (1988a) found that with phalaris, MCP was four times as effective per unit of P as dicalcium phosphate (DCP), which in turn was approximately four times as effective as tricalcium phosphate (TCP). The most soluble source was also found to be the most damaging to emergence.

Despite the marked differences in effectiveness, the relative efficiencies were not directly related to their solubilities because MCP is 50 times as soluble as DCP, which is 16 times as soluble as TCP (Scott and Blair, 1988a).

Hooke (1990) found that low rates of P coated on Verano stylo were as effective as 20 kg P/ha broadcast with the seed. Scott and Blair (1988b) found that MCP at 5 kg P/ha on phalaris was as effective as 20 kg P/ha drilled. Plant height was also less variable, suggesting that the nutrients in the coating were more uniformly available. Scott, Hill, and Jessop (1991) found that wheat coated with phosphorus (P) at 5 kg P/ha had a shoot P content equivalent to that supplied with drilled P at 7 kg P/ha, whereas the same coating resulted in root growth equivalent to a drilled application of 40 kg P/ha. They also reported that the phosphorus-sorbing capacity of soil had relatively little impact on the efficacy of coated vs. drilled phosphorus fertilizer treatments.

When similar coatings were tested on wheat and barley in the field by Ascher (1994), significant differences could not be detected, suggesting perhaps that glasshouse conditions (with the necessarily frequent watering of pots) may provide conditions more suited to the observation of differences.

Sulfur. Sulfur is most commonly applied as gypsum, although elemental sulfur has also been applied without damage to seeds. For example, seeds of groundnut (*Arachis hypogaea*) were found to be tolerant of gypsum coatings at a rate of 12 g/kg seed (Bhaskar and Shankar, 1993). Gilbert and Shaw (1979) found that sulfur could be successfully applied to seeds of *Stylosanthes* spp. for surface sowing.

Sulfur has been frequently incorporated in nutrient coatings with other elements (see below).

Nitrogen. Scott et al. (1987) showed urea coatings to be damaging to wheat germination but the level of damage could be alleviated substantially by including a urease inhibitor, phenylphosphorodiamidate (PPD), in the coating at 1% of the weight of urea; this suggested that slowing urea hydrolysis reduced injury and that free ammonia was the main toxic agent. PPD was more effective than bentonite additions in reducing injury, although bentonite with pH of 5 reduced injury slightly more than one of pH 9, presumably due to less formation of free ammonia.

As with other nutrients, slowly available forms of nitrogen such as isobutylidene diurea (IBDU) are safe to emerging seeds but appear to be ineffective in promoting growth (Scott, Mitchell, and Blair, 1985).

Phosphorus and Nitrogen. Scott, Archie, and Hay (1992) found that pellets made using perennial ryegrass with a N:P:seed ratio of 1:1:1 had the best effect on establishment. Rebafka, Batione, and Manschnen (1993) found that P and N supplied as ammonium dihydrogen phosphate was the most successful nutrient seed coating for pearl millet (*Pennisetum glaucum*); this was attributed to the enhancement of P uptake by ammonium ions. Although seedling emergence was reduced when rates were greater than 0.5 mg/seed, this fertilizer was found to be superior as a coating to others tested (superphosphate, monocalcium phosphate, sodium dihydrogen phosphate and sodium triphosphate). When this seed was sown in acid, P-deficient sandy soil, they observed an increase in seedling yield at 20 days by 280%, P content by 330%, total biomass at maturity by 30%, and grain yield by 45%. It is noteworthy that the size of the effect was greatest in seedlings and diminished at later growth stages.

Scott et al. (1987) found that a mixture of calcium nitrate and MCP was quite damaging, although neither MCP nor calcium nitrate were particularly damaging alone. Garrote et al. (1990b) found that a combination of urea and MCP could be successfully applied to seeds of tall fescue without significant seed damage and yet resulting in significant enhancement of seedling growth in the field.

Phosphorus and Sulfur. Many references cite the damage caused to seeds by superphosphate and yet, as Rader, White, and Whittaker (1943) point out, it is considered a relatively "neutral" fertilizer in terms of its salt index. Cameron (1994) found that combinations of soluble sources of sulfur and phosphorus (such as found in superphosphate) appear to be particularly damaging in spite of the fact that neither monocalcium phosphate nor gypsum cause severe damage by themselves. Further more detailed investigations led to the finding that phosphorus uptake by the germinating seed was enhanced by the presence of sulfur, thus leading to toxic levels in the seeds. For this reason, many coating formulations use reverted superphosphate to avoid this kind of damage to the germinating seed. This is in spite of the fact that Scott, Mitchell, and Blair (1985) found that reverted superphosphate was quite inadequate in its supply of P to ryegrass seedlings, although it was satisfactory in supplying S. In that same experiment, partially acidulated rock phosphate containing elemental sulfur contributed significant amounts of P and was also effective in providing S to establishing ryegrass seedlings.

Phosphorus, Sulfur and Nitrogen. Combinations of phosphorus, sulfur and nitrogen have been found to be safe to germinating seeds and effective in promoting seedling growth in both the glasshouse and field (Cameron, Scott, and White, 1993). It appears that the combination of the three elements overcomes some of the toxicity problems associated with coatings containing just two elements. Although the coating led to significant increases in plant growth, emergence was nevertheless slightly delayed and comparisons with drilled applications of fertilizer were not as favourable in field experiments as in glasshouse trials. Nevertheless, as shown in Figure 3, the coated seeds grew at least as well as drilled applications and far better than those seedlings which had no fertilizer applied.

Trace Elements. Coatings containing trace elements have been described in the literature for molybdenum, iron, zinc, manganese and boron. The application of iron, zinc and manganese is of great importance in alkaline soils where the availability of these elements is decreased. Molybdenum is commonly applied to legume seed with lime, especially when sown on acid soils. A range of chelated and mineral forms of trace elements have been used in seed coatings.

Zinc coated onto rice seed before planting resulted in yield increases (see Scott, 1989a). Gangwar and Singh (1994) found that zinc oxide coating on lentils (at 0.1% of seed weight) produced a greater increase in yield on a sandy loam soil than foliar or soil applications of zinc sulfate. The zinc oxide coating also enhanced nitrogen and phosphorus uptake, whilst zinc sulfate applications lowered P uptake.

FIGURE 3. Dry matter yield of tall fescue seedlings 5 months after sowing with either banded or coated nutrients (MCP = monocalcium phosphate; N = urea; S = gypsum).

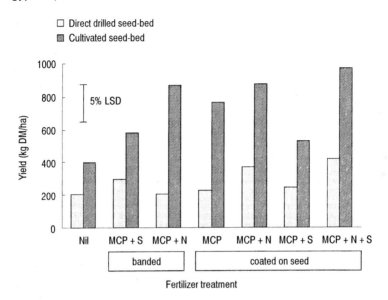

Work by Ascher (1994) has shown how, in high pH soils, seed coatings containing manganese as manganese sulfate or manganese dextrolac can contribute relevant quantities of manganese to cereal crops, at times having significant effects through to the final yield. Combinations of the manganese seed coating with the fungicide mancozeb further enhanced performance.

Safety of Seed Nutrient Coating

The value of nutrient coated seeds in practice was questioned by Smid and Bates (1971) who stated that, if sufficient fertilizer in coatings were to become possible, a way had to be found to reduce the toxicity of fertilizers to germinating seeds.

Whilst the aim has been to improve seed performance, some of the treatments imposed have had negative consequences. There are numerous references of damage to seeds and seedlings due to seed coatings containing nutrients; even today, effective nutrient coatings with reliable performance and wide application are still to be developed.

In spite of there being numerous negative reports of nutrient seed coatings, there are enough positive reports to suggest that the technology is capable of being improved sufficiently to be adopted more widely and routinely. In addition to those cited above, other positive reports include:

- Growth promotion of *Stylosanthes guianensis* with sulfur coatings containing gypsum or elemental sulfur (Gilbert and Shaw, 1979),
- Dicalcium phosphate, urea and sulfate of potash increased the growth of tall fescue and Kentucky bluegrass (see Scott, 1989a),
- Urea, calcium phosphate and KCl were successful in enhancing the growth of Italian ryegrass (Yano, 1973), and
- Low rates of soluble P promoting the growth of buffel grass (Silcock and Smith, 1982).

Methods of Improving Coating Safety

Protective Coatings

Protective coatings were evaluated by Smid and Bates (1971) who found sucrose and polyvinyl-acetate gave some protection, but it was insufficient to be of practical importance. Scott (1986) also found a polymer coating of tung oil on lucerne seed gave some protection from phosphorus fertilizer injury.

Scott (1989b) showed that increasing coatings of polymer reduced injury to emerging lucerne coated with MCP. This was shown to be due to the polymer simulating the effect of the lemma/palea of grasses by slowing down emergence and thereby delaying contact between the emerging root and possibly toxic levels of fertilizer.

Reducing Fertiliser Solubility

Using less soluble fertilizers is one way that can be used to make coatings less damaging to seeds (Scott and Hay, 1974). However, as found by Scott and Blair (1988a, b), as solubility decreased, so did efficacy.

In common with many other authors, Hayward and Scott (1993) found that lime-reverted superphosphate was the safest nutrient coating for emerging seedlings but, as noted above, the supply of P to seedlings is minimal.

Rock phosphate sources with a wide range of reactivities have been investigated as candidates for seed coating. This work has clearly shown that the phosphorus contained in these sources is poorly available to establishing seedlings (Cameron and Scott, unpublished).

Controlled-Release Nutrient Coatings

Slowing the release rate of fertilizers may lead to greater efficiency of use of fertilizers (Kydonieus, 1980). Lawton (1961) showed that, when sown near controlled release, coated fertilizer, wheat seedlings emerged with less damage than when sown near uncoated fertilizer.

Little research has been done on controlled release of fertilizers for seed coatings. Whilst it appears that seeds require soluble nutrients to stimulate growth, they need to be released slowly, especially during germination. Such release rates can be engineered through controlled release technology but, to the author's knowledge, this has not been successfully implemented yet with seed coatings.

Controlled release has the potential to provide a more even distribution of nutrients over an extended growth period. It is also possible that their release rate can be tailored to satisfy the changing demands of the plant over time.   '

In the case of micronutrients, Mortvedt (1994) suggested that, because of the high residual effects of soluble sources of micronutrients such as copper and zinc, controlled release may not be necessary. Iron and manganese can become unavailable to plants due to oxidation in the soil while soluble boron fertilizers can be leached somewhat, but not as much as nitrate. Controlled release of molybdenum is not needed as only trace quantities are required for plant growth.

Better Chemical Formulations

Work by Cameron, Scott, and White (1993) has shown that it is possible to combine at least three elements together in a safe seed-coating formulation. However, a more detailed understanding of the interactions between the component phosphorus, nitrogen, and sulfur, with the seed and the soil is needed to improve the performance of such coatings sufficiently to enable widespread commercialisation. The key to creating better formulations is to understand the chemistry of nutrient release and sorption with the soil that will allow germination and emergence to proceed unhindered and yet leave the nutrients relatively available in the soil surrounding the seed.

Fertiliser-Rhizobia Compatibility

Faizah, Broughton, and John (1980) found that relatively insoluble fertilizers were compatible with rhizobia, but suggested that soluble fertilizers were damaging. Scott and Archie (1978) found that molybdenum at high rates is toxic to rhizobia. Lowther and Johnstone (1979) showed lime to be much safer to rhizobia than rock phosphate, thermophos or gypsum. It is unlikely that rhizobia could be applied to legume seeds with soluble fertilizer coatings without some form of sophisticated protection for the rhizobia.

Efficacy

The efficacy of nutrient coatings has been shown to be superior to other forms of fertilizing by a number of researchers. These include comparisons with broadcast nutrients (Gilbert and Shaw, 1979; Scott and Blair, 1988b; Hooke, 1990) and drilled fertiliser (Smid and Bates, 1971; Scott and Blair, 1988b; Scott, Hill, and Jessop, 1991). Hooke (1990) showed that is was possible to enhance seedling growth rate of Verano stylo by applying a soluble form of phosphorus in a seed coating. Seedling growth from the coated treatment containing 4.6 kg P/ha increased by 600% over that of the nil P control, whereas a comparable commercial coating (supposedly containing P) reduced growth by 23%. This coating also compared favourably with a broadcast application of 20 kg P/ha which increased seedling growth by 280% over the control.

Nutrient coatings can result in less fertilizer contacting the soil and hence greater availability of applied nutrients (Smid and Bates, 1971). They showed that, at low rates on corn, nutrient coatings were 3-4 times as effective in promoting early seedling growth as banded fertilizer.

Using ^{32}P-labelled fertilizer coatings, Scott, Rocks, and Blair (1987) found that, regardless of soil P level, the coated P fertilizer was taken up more readily from coatings than from drilled granules, presumably due to its greater proximity to the seedlings.

In early establishment, seedlings require a highly available source of nutrients. It has been shown that seedling growth can be promoted by nutrient coatings even in soils high in plant available P (Richards, Bates, and Sheppard, 1985). The efficacy from seed coatings may be higher under cool conditions where root growth rate is restricted (Robinson, Sprague, and Gross, 1959).

Storage Conditions

Scott et al. (1987) found that wheat seed coated with urea and stored in moisture-proof bags for 11 months had the same emergence as when freshly coated. The key to successful storage of coated seed is to dry the seed adequately and then to seal it in vapour-proof packaging. In a storage study by Cameron, White, and Scott (unpublished results), the dominant factors leading to seed damage during storage were, in decreasing order, temperature (warm was more damaging), humidity (high was more damaging) and then bag type (cloth was worse than plastic than foil).

COMMERCIAL SEED COATING

Many of the reports relating to seed coatings relate to studies on commercial coatings which often have undisclosed formulations. Details of their exact chemical composition and particle size are often not given. This results in some confusion in the literature, as the reason for performance at times cannot be ascribed to a particular cause. Coated seeds have been commercialised to the greatest extent in high value vegetable industries (e.g., lettuce) and in specialist crops such as sugarbeets. Also pre-inoculated seeds with a coating protecting the *Rhizobium* inoculant have been commercialised for many years (e.g., Coated Seeds Ltd., 1975). Whilst coatings containing fertilizers have been marketed for some years (especially those for the turf grass industries), most still depend on relatively insoluble forms of nutrients which provide relatively little boost to seedling growth.

Ultimately, successful commercial nutrient coatings will be mass produced when they provide demonstrable benefits in the field. This will also be aided by existence of economies of scale to make the price of the coated seed affordable, and an adequate quality assurance program which will guarantee performance of the coated seeds.

CONCLUSIONS AND FUTURE DIRECTIONS FOR RESEARCH

If nutrient coated seeds are to become a widespread reality, it is crucial that the vulnerability of the seed as a living organ is recognised by those formulating the coatings. Coatings must not adversely affect the germinating seed and yet, soon after germination is complete, they must release readily available forms of nutrients.

As pointed out by Scott (1993) when writing about models of coated seeds, there is a need to move beyond "trial and error" studies with seed coatings towards a more mechanistic understanding. Whilst there is ample evidence that fertilizers can be delivered to establishing plants via seed coatings, there remains much to be learned about interactions both between components of complex coatings and between the coating, the seed and its environment.

Some of the areas considered worthy of investigation include the role of magnesium salts in overcoming nutrient toxicity, combinations of nutrient coating and soaking, and more detailed investigations of controlled release formulations. Also of interest is the area of complex multi-nutrient formulations with additions of protectants and growth promoters. The separation of any priming effects from those of soaking in nutrients also needs investigation.

REFERENCES

Aghinoni, I. and S.A. Barber. (1980). Predicting the most efficient placement method for corn. *Soil Science Society of America Journal* 44: 1016-1020.

Ascher, J.S. (1994). Phosphorus and manganese seed coatings for crop growth and yield. PhD thesis, University of New England, Armidale, Australia. 169 p.

Ascher, J.S., J.M. Scott, and R.S. Jessop. (1987). Tolerance of a range of crop species to seed coating with monocalcium phosphate. *Proceedings of 4th Australian Agronomy Conference*, Melbourne, Australia. p. 239.

Ascher, J.S., J.M. Scott, and R.S. Jessop. (1990). Effect of soil texture on injury to crop species by seed coating with monocalcium phosphate. *Proceedings of 5th Australian Agronomy Conference*, Perth, Australia. p. 532a.

Barber, S.A. and J.L. Kovar. (1985). Review: Principles of applying phosphorus fertiliser for greatest efficiency. *Journal of Fertiliser Issues* 2: 91-94.

Bhaskar, S. and K.S. Shankar. (1993). Productivity and economics of groundnut (*Arachis hypogaea*) in rice (*Oryza sativa*) fallows as influenced by gypsum and sulphur. *Indian Journal of Agronomy* 38: 53-55.

Bolland, M.D.A. and Baker, M.J. (1988). High phosphorus concentrations in seed of wheat and annual medic are related to higher rates of dry matter production of seedlings and plants. *Australian Journal of Experimental Agriculture* 28: 765-770.

Bose, B. and T. Mishra. (1992). Response of wheat seeds to pre-sowing seed treatment with Mg(NO₃)₂. *Annals of Agricultural Research* 13: 132-136.

Cameron, D.B. (1994). An examination of the use of nutrient and fungicide seed coatings for pasture establishment. M.Rur.Sc. thesis. University of New England, Armidale, Australia. 136 p.

Cameron, D.B., J.M. Scott, and A.V. White. (1993). Effect of phosphorus, nitrogen, sulphur and fungicide seed coatings on the emergence and dry matter production of tall fescue and red clover. *Proceedings of 7th Australian Agronomy Conference*, Adelaide, Australia. p. 415.

Carter, O.G. (1969). The effects of fertilisers on germination and establishment of pastures and fodder crops. *Wool Technology and Sheep Breeding*, July: 67-75.

Chhipa, B.R. and P. Lal. (1993). Ionic ratios as the basis of salt tolerance in wheat. *Agrochimica* 37: 63-67.

Coated Seeds Ltd. (1975). New Zealand Patent No. 164536.

Collis-George, N. (1987). Effects of soil physical factors on imbibition, germination, root elongation and shoot movement. In *Crop Establishment Problems in Queensland*, eds. I.M. Wood, W.H. Hazard, and F.R. From, Occasional Publication 34. Brisbane, Australia: Australian Institute of Agricultural Science, pp. 23-41.

Dowling, P.M., R.J. Clements, and J.R. McWilliam. (1971). Establishment and survival of pasture species from seeds sown on the soil surface. *Australian Journal of Agricultural Research* 22: 61-74.

Ensminger, L.E., J.T. Hood, and C.T. Willis. (1965). The mechanism of ammonium phosphate injury to seeds. *Proceedings of Soil Science Society of America* 29: 320-327.

Faizah, A.W., W.J. Broughton, and C.K. John. (1980). Rhizobia in tropical legumes. II. Survival in the seed environment. *Soil Biology and Biochemistry* 12: 219-227.

Funsten, S.R. and F.W. Burgesser. (1951). U.S. Pat No. 2,579,732 (25 Dec).

Gangwar, K.S. and N.P. Singh. (1994). Studies on zinc nutrition on lentil in relation to dry matter accumulation, yield and N, P uptake. *Indian Journal of Pulses Research* 7: 33-35.

Garrote, B.P., J.M. Scott, and G.J. Blair. (1987). Effect of seed structure on the tolerance during emergence of five grasses to phosphorus seed coating. *Proceedings of 4th Australian Agronomy Conference*, Melbourne, Australia. p. 248.

Garrote, B.P., J.M. Scott, P.W.G. Sale, and G.J. Blair. (1990a). Effect of seed structure on the water relations of germinating pasture grass seeds. *Proceedings of 5th Australian Agronomy Conference*, Perth, Australia. p. 444.

Garrote, B.P., J.M. Scott, P.W.G. Sale, and G.J. Blair. (1990b). Effect of nitrogen and phosphorus seed coating on the emergence of tall fescue. *Proceedings of 5th Australian Agronomy Conference*, Perth, Australia. p. 543.

Gilbert, M.A. and K.A. Shaw. (1979). A comparison of sulfur fertilizers and sulfur seed pellets on a *Stylosanthes guianensis* pasture on a euchrozem in north Queensland. *Australian Journal of Experimental Agriculture and Animal Husbandry* 19: 241-246.

Guttay, J.R. (1957). The effect of fertilizer on the germination of wheat and oats. *Michigan Agricultural Experiment Station Quarterly Bulletin* 40: 193-202.

Guttay, J.R., J.A. Stritzel, A.J. Englehorn, and C.A. Black. (1957). Treatment of corn seed with phosphate. *Agronomy Journal* 49: 98-101.

Hayward, G.D. and W.R. Scott. (1993). The effect of fertilizer type on brassica establishment and yield. *Proceedings of the Annual Conference of the Agronomy Society of New Zealand* 23: 35-41.

Heydecker, W. and P. Coolbear. (1977). Seed treatments for improved performance– survey and attempted prognosis. *Seed Science and Technology* 5: 353-425.

Hoeft, R.G., L.M. Walsh, and E.A. Liegel. (1975). Effect of seed placed fertilizer on the emergence (germination) of soybeans (*Glycine max* L.) and snapbeans (*Phaseolus vulgaris* L.). *Communications in Soil Science and Plant Analysis* 6: 655-664.

Hooke, J.J. (1990). The establishment of *Stylosanthes*-based pastures in Northern Australia, with special reference to the fertiliser practice adopted. B. Rur. Sc. Thesis. University of New England, Armidale, Australia.

Kurdikeri, M.B., B. Aswathaiah, and S.R. Prasad. (1993). Seed invigoration studies in maize hybrids. *Seed Research* 21: 8-12.

Kydonieus, A.F. (1980). Fundamental concepts of controlled release. In *Controlled Release Technologies: Methods, Theory and Applications*, ed. A.F. Kydonieus. Boca Raton, Florida, USA: CRC Press, pp. 1-19.

Lawton, K. (1961). The effect of coatings on the dissolution of fertilisers and the uptake of fertiliser potassium by plants. *Journal of Agricultural and Food Chemistry* 9: 276-280.

Liu, L.X. and J.D. Litster. (1993a). Coating mass distribution from a spouted bed seed coater. *Powder Technology* 74: 259-270.

Liu, L.X. and J.D. Litster. (1993b). Spouted bed coating: the effect of process parameters on maximum coating rate and elutriation. *Powder Technology* 74: 215-230.

Longnecker, N.E., N.E. Marcar, and R.D. Graham. (1991). Increased manganese content of barley seeds can increase grain yield in manganese-deficient conditions. *Australian Journal of Agricultural Research* 42: 1065-1074.

Lowther, W.L. and P.D. Johnstone. (1979). Coating materials for commercial inoculated and coated clover seed. *New Zealand Journal of Agricultural Research* 22: 475-478.

McEvoy, J., J.S. Ascher, R.D. Graham, and P. Hardy. (1988). Seed coating with manganese sulphate is a method of reducing manganese deficiency and improving early seedling vigour. In *International Symposium on Manganese in Soils and Plants: Contributed Papers*, eds. M.J. Webb, R.O. Nable, R.D. Graham, and R.J. Hannam. Adelaide, Australia: Manganese Symposium 1988 Inc., pp. 77-79.

Miller, M.H., T.E. Bates, D. Singh, and A.S. Baweja. (1971). Response of corn to small amounts of fertilizer placed with the seed: I. Greenhouse studies. *Agronomy Journal* 63: 365-368.

Mortvedt, J.J. (1994). Needs for controlled-availability micronutrient fertilizers. *Fertilizer Research* 38: 213-221.

Olson, R.A., and A.F. Dreier. (1956). Fertiliser placement for small grains in relation to crop stand and nutrient efficiency in Nebraska. *Soil Science Society of America Proceedings* 20: 19-24.

Paul, S.R., A.K. Choudhury, and S.C. Dey. (1993). Effect of pre-sowing seed hardening with potassium salts and levels of applied potassium on growth and yield of wheat under rainfed conditions. *Journal of Potassium Research* 9: 160-165.

Philip, J.R. (1958). The osmotic cell, solute diffusibility, and the plant water economy. *Plant Physiology* 33: 264-271.

Rader, L.F., L.M. White, and C.W. Whittaker. (1943). The salt index–a measure of the effect of fertilizers on the concentration of the soil solution. *Soil Science* 55: 201-218.

Rebafka, F.P., A. Bationo, and H. Marschner. (1993). Phosphorus seed coating increases phosphorus uptake, early growth and yield of pearl millet (*Pennisetum glaucum* (L.) R. Br.) grown on an acid sandy soil in Niger, West Africa. *Fertilizer Research* 35: 151-160.

Richards, J.E., T.E. Bates, and S.C. Sheppard. (1985). The effect of broadcast P applications and small amounts of fertilizer placed with the seed on continuously cropped corn (*Zea mays* L.). *Fertilizer Research* 6: 269-277.

Robinson, R.R., V.G. Sprague, and C.F. Gross. (1959). The relation of temperature and phosphate placement to growth of clover. *Soil Science Society of America Proceedings* 23: 225-228.

Scott, D. (1975). Effects of seed coating on establishment. *New Zealand Journal of Agricultural Research* 18: 59-67.

Scott, D. and W.J. Archie. (1978). Sulphur, phosphate, and molybdenum coating of legume seed. *New Zealand Journal of Agricultural Research* 21: 643-649.

Scott, D. and R.J.M. Hay. (1974). Some physical and nutritional effects of seed coating. *Proceedings of 12th International Grasslands Congress*, Moscow, Russia, Volume I, Part II, pp. 523-531.

Scott, D., W.J. Archie, and R.J.M. Hay. (1992). Effect of sowing method, soil moisture and temperature on establishment from coated grass seed. *Proceedings of the New Zealand Grassland Association* 54: 131-133.

Scott, J.M. (1986). Seed coating as an aid to pasture establishment. Ph.D. thesis, University of New England, Armidale, Australia. 203 p.

Scott, J.M. (1989a). Seed coatings and treatments and their effects on plant establishment. *Advances in Agronomy* 42: 43-83.

Scott, J.M. (1989b). Developing effective nutrient seed coatings for grasses and legumes. *Proceedings of 16th International Grassland Congress*, Nice, France, pp. 91-92.

Scott, J.M. (1993). Towards an ideal seed coating. *Proceedings of 17th International Grassland Congress*, New Zealand. pp. 1877-1878.

Scott, J.M. and G.J. Blair. (1988a). Phosphorus seed coatings for pasture species. I. Effect of source and rate of phosphorus on emergence and early growth of phalaris (*Phalaris aquatica* L.) and lucerne (*Medicago sativa* L.). *Australian Journal of Agricultural Research* 39: 437-445.

Scott, J.M. and G.J. Blair. (1988b). Phosphorus seed coatings for pasture species. II. Comparison of effectiveness of phosphorus applied as seed coatings, drilled or broadcast applications in promoting early growth of phalaris (*Phalaris aquatica* L.)

and lucerne (*Medicago sativa* L.). *Australian Journal of Agricultural Research* 39: 447-456.

Scott, J.M. and L.M. O'Donnell. (1987). The susceptibility of lucerne to soluble phosphate during germination. *Proceedings of 4th Australian Agronomy Conference*, Melbourne, Australia. p. 247.

Scott, J.M., G.J. Blair, and A.C. Andrews. (1997). The mechanics of coating seeds in a small rotating drum. *Seed Science and Technology* (In Press).

Scott, J.M., C.B. Hill, and R.S. Jessop. (1991). Growth chamber study of phosphorus applied as drilled granules or as seed coatings to wheat sown in soils differing in P-sorption capacity. *Fertilizer Research* 29: 281-287.

Scott, J.M., R.S. Jessop, R.J. Steer, and G.D. McLachlan. (1987). Effect of nutrient seed coating on the emergence of wheat and oats. *Fertilizer Research* 14: 205-217.

Scott, J.M., C.J.M. Mitchell, and G.J. Blair. (1985). The effect of nutrient seed coating on the emergence and early growth of perennial ryegrass. *Australian Journal of Agricultural Research* 36: 221-231.

Scott, J.M., R.L. Rocks, and G.J. Blair. (1987). Phosphorus uptake by phalaris from ^{32}P-labelled seed coatings and drilled granules. *Proceedings of 4th Australian Agronomy Conference*, Melbourne, Australia. p. 273.

Sharples, G.C. and J.P. Gentry. (1980). Lettuce emergence from vermiculite seed tablets containing activated carbon and phosphorus. *HortScience* 15: 73-75.

Silcock, R.G. and F.T. Smith. (1982). Seed coating and localised application of phosphate for improving seedling growth of grasses on acid, sandy red earths. *Australian Journal of Agricultural Research* 33: 785-802.

Smid, A.E. and T.E. Bates. (1971). Response of corn to small amounts of fertilizer placed with the seed. V. Seed coating compared with banding. *Agronomy Journal* 63: 380-384.

Vartha, E.W. and P.T.P. Clifford. (1973). Effect of seed coating on establishment and survival of grasses, surface-sown on tussock grasslands. *New Zealand Journal of Experimental Agriculture* 1: 39-43.

Welch, L.F., D.L. Mulvaney, L.V. Boone, G.E. McKibben, and J.W. Pendeleton. (1966). Relative efficiency of broadcast versus banded phosphorus for corn. *Agronomy Journal* 58: 283-287.

Wilhelm, N.S., R.D. Graham, and A.D. Rovira. (1988). Application of different sources of manganese sulphate decreases take-all (*Gaeunannomyces graminis* vao *tritici*) of wheat grown in a manganese deficient soil. *Australian Journal of Agricultural Research* 38: 1-10.

Wythes, M. (1990). Phosphorus seed coating and its influence on the competitiveness of phalaris sown into weeds. B.Rur.Sc. Honours thesis, University of New England, Armidale, Australia. 85 p.

Yano, A. (1973). [Studies on the improvement of vegetation on sloping land in warmer districts. 9. The method of pelleting Italian ryegrass seed.] (In Japanese, English summary). *Journal of Japanese Society of Grasslands Science* 19: 269-275.

SUBMITTED: 04/22/97
ACCEPTED: 07/16/97

Nutrient Management, Cultivar Development and Selection Strategies to Optimize Water Use Efficiency

Jessica G. Davis
James S. Quick

SUMMARY. Cultivars can be selected for traits that improve water use efficiency, and fertilizers and other nutrient sources also can be managed to optimize water use efficiency of a crop. Both cultivar selection and nutrient management impact water use efficiency by altering photosynthetic rate, yield, rooting characteristics, transpiration, or soil evaporation. In order to optimize water use efficiency, cultivar and nutrient decisions should be made jointly. This integrated approach will lead to improvements in water use efficiency, with an increase in productivity and profitability per unit water. *[Article copies available for a fee from The Haworth Document Delivery Service: 1-800-342-9678. E-mail address: getinfo@haworthpressinc.com]*

KEYWORDS. Cultivar selection, evapotranspiration, roots, water use efficiency, yield

---

Jessica G. Davis, Associate Professor and Extension Environmental Soil Scientist; and James S. Quick, Professor and Wheat Breeder, Department of Soil and Crop Sciences, Colorado State University, Fort Collins, CO 80523-1170 USA.

Address correspondence to: Jessica G. Davis, Department of Soil and Crop Sciences, Colorado State University, Fort Collins, CO 80523-1170 USA (E-mail: jgdavis@ lamar.colostate.edu).

[Haworth co-indexing entry note]: "Nutrient Management, Cultivar Development and Selection Strategies to Optimize Water Use Efficiency." Davis, Jessica G., and James S. Quick. Co-published simultaneously in *Journal of Crop Production* (Food Products Press, an imprint of The Haworth Press, Inc.) Vol. 1, No. 2 (#2), 1998, pp. 221-240; and: *Nutrient Use in Crop Production* (ed: Zdenko Rengel) Food Products Press, an imprint of The Haworth Press, Inc., 1998, pp. 221-240. Single or multiple copies of this article are available for a fee from The Haworth Document Delivery Service [1-800-342-9678, 9:00 a.m. - 5:00 p.m. (EST). E-mail address: getinfo@haworthpressinc.com].

WATER USE EFFICIENCY

Nutrient management has recently broadened its focus from a primary goal of yield or profit maximization to include an equally important goal of minimization of nutrient leaching and runoff losses. In addition to nutrient management impacts on water quality, nutrients also can be administered to conserve water quantity by optimizing water use efficiency. Over 30 years ago, Frank G. Viets, Jr. authored a review of the influence of fertilizers on water use efficiency (Viets, 1962). The current emphasis on agricultural impacts on water quality makes it imperative to understand nutrient interactions with all other inputs and losses, particularly those which influence conservation of water quantity and quality.

This chapter will address the potential for cultivar selection, nutrient management, and their interactions to bring about improved water use efficiencies. Changing crop species, cropping systems, or soil physical properties to optimize water use efficiency is outside the scope of this chapter. Water use efficiency (WUE) can be defined as: WUE = Y/ET, where Y = yield of grain or biomass, and ET = evapotranspiration. WUE can also be defined as photosynthesis/transpiration, which is commonly termed transpiration efficiency. Nutrients and cultivars can affect WUE by altering yield, photosynthesis, evaporation, or transpiration.

PLANT CHARACTERISTICS FOR USE
IN CULTIVAR DEVELOPMENT

Improved performance of crop cultivars under water-limited conditions, i.e., drought tolerance, involves selection for more than just WUE. Water use efficiency itself is not likely to be altered genetically, but may increase as plants are altered for drought resistance. WUE is an end point, and like grain yield, will only be altered by genetic advances for underlying mechanisms. Plants showing improved growth or productivity with limited water are considered drought tolerant regardless of how the improvement occurs or whether the WUE is affected (Boyer, 1996). Early maturity is a mechanism of drought escape which effectively allows the plant to avoid periods of low water availability at the end of the season. These mechanisms may improve WUE by increasing yields under dry conditions. Dehydration avoidance and dehydration tolerance are considered the major general categories of drought resistance that are evaluated when searching for plant traits that contribute to improved performance under drought (Blum, 1988; Boyer, 1996). Water use efficiency is simply the end result of many complex plant reactions, much like grain yield, and needs to be considered from basic physiological and morphological viewpoints in order to provide possible paths for improvement. During the past 20 years, there has been a general increase in yields of modern crops with little change in total above-ground biomass. Harvest index (HI) in crops (ratio of economically valuable portion to total biomass) is very closely related to crop improvement through breeding for drought resistance and WUE since grain yield is the major economic product.

Despite the above limitations, sustained attempts to indirectly improve WUE are required since any increase may have far reaching effects in dry environments (Richards, 1987). There are numerous traits that could conceivably increase WUE, via their effects on transpiration and/or carbohydrate translocation, if selected. They can be separated into (1) simple morphological traits that are easily manipulatable, such as the presence of awns, pubescence and glaucousness, (2) more complexly inherited traits such as carbon isotope discrimination (CID), specific leaf weight, and abscisic acid, and (3) traits that at first appear unrelated to WUE, such as cooler leaf temperature, seedling vigor, and dry matter partitioning to roots.

Recently, two techniques have been developed that enable screening for WUE and HI in breeding populations of wheat and barley (Turner et al., 1989). The two techniques are (1) the use of stable CID for the assessment of transpiration efficiency and (2) the use of a desiccation agent to assess stable HI via grain-filling ability under conditions of post-anthesis water shortage. The results with CID have been mixed; in greenhouse studies with wheat, cotton, barley, sunflower, and peanuts, a negative correlation between WUE and CID was observed and can be used to distinguish differences among genotypes in WUE. Field studies in Australia, Syria, and elsewhere as reviewed by Morgan et al. (1993) indicate that total dry matter and grain yield were positively, negatively, or not correlated with CID values in wheat and barley. The desiccation technique involves the application of a desiccant 10 to 15 days after anthesis, which arrests photosynthesis and allows the estimation of variation in the transfer of assimilates from the stem, leaves, and roots to the spike. The extent of reduction in grain size of treated compared with untreated irrigated plants is considered to be a measure of the plant's ability to transfer assimilates to the grain under late-season drought. The method has been shown to correlate with the reduction in grain size arising from drought and to be reliable from season to season, from site to site, and among crosses (Blum, 1988; Haley and Quick, 1993).

Another important trait affecting WUE by reducing transpiration is leaf glaucousness, the presence of a waxy cuticle on the leaf surface. In wheat, near-isogenic glaucous lines have produced higher grain yield than non-glaucous lines when grown in water-limiting environments (Richards, 1987). Glaucous lines have (1) lower transpiration at night, (2) lower transpiration for a given rate of photosynthesis during the day, and (3) cooler photosynthetic surfaces, presumably due to increased light reflectance.

Several studies have demonstrated the importance of a deep, extensive root system in effective water extraction, and efforts have been made to increase the root system in many crop plants. Also, increased root hydraulic resistance was proposed by Richards and Passioura (1981) as a way to conserve stored soil water for grain filling. Decreasing the diameter of the main xylem vessel in the seminal roots of wheat has resulted in increased yield under very dry conditions in Australia. There has been no yield penalty under higher rainfall conditions since the crown (secondary) root system overrides the effect of the small xylem vessel.

Osmotic adjustment has also provided a physiological opportunity to improve WUE. Morgan (1983) selected wheat lines with superior osmotic adjustment under dry conditions and observed improved yields in wheat. The improved yields were at no cost to yield under optimum conditions. This probably is explained by the low metabolic cost of osmotic adjustment together with the lack of osmotic adjustment under optimum conditions (Boyer, 1996).

Fischer, Edmeades, and Johnson (1989) selected a population of tropical maize for several physiological attributes likely to improve drought tolerance, and found that most of these traits conferred a yield gain; however, early silking relative to pollen shed generally was the most important. They were able to improve yield by as much as 410 kg ha $^{-1}$ under severe dehydration, and not at the expense of yield in hydrated conditions.

Since the expression of most traits of interest in breeding for drought resistance is strongly influenced by the environment and are quantitatively inherited, the use of molecular marker-assisted selection (MAS) would be very helpful (Martin et al., 1989). Martin et al. (1989) have identified three RFLP (restricted fragment length polymorphism) loci associated with C isotope discrimination in tomato. Mian et al. (1996) used RFLP analyses in soybeans to associate QTL's (quantitative trait loci) with WUE and leaf ash (the latter being generally related to WUE). They found a total of four and six independent RFLP markers associated with WUE and leaf ash, and if combined, each group of markers would explain 38 and 53% of the variability in the respective traits. Schneider, Brothers, and Kelly (1997) evaluated two common bean *(Phaseolus vulgaris* L.) populations under stress and nonstress conditions to identify RAPD (random amplified polymorphic DNA) markers associated with drought resistance. They found four RAPD markers in one population and five in another that were consistently associated with yield under stress, yield under nonstress, and/or geometric mean yield across a broad range of environments. Marker-assisted selection under selected stress environments with one population showed that MAS improved performance 11% under stress and 8% under non-stress conditions, whereas conventional selection based on yield performance failed to increase performance. In the second population, response to conventional selection was three times greater than that to MAS. Genetic markers accounted for a similar level of variation in both crosses; however, the heritability for yield itself in the second population was three times greater than in the first. These results support quantitative genetic theory that the effectiveness of MAS is inversely proportional to the heritability of the trait under examination (Lande and Thompson, 1990). Future MAS studies hold great promise for improving drought resistance since environmental influences in the field severely limit conventional selection progress. However, further research is necessary to determine if markers are population specific, to identify additional polymorphisms, and to determine the most appropriate testing environments.

Regardless of the trait considered for improvement, and its relationship with WUE through improved drought resistance, there must be adequate genetic variation, reasonably high heritability, and/or effective MAS for the breeder to make significant selection progress. These traits must also be validated by developing and testing contrasting lines under the conditions of interest. These lines can be developed by random line selection within a breeders materials or by backcrossing if the trait is simply inherited. Cooperation among the breeders, geneticists, and physiologists is essential to determine priorities, assess the impact of association with other traits, and to share ideas and cost of development.

PATHWAYS FOR NUTRIENT IMPACTS ON YIELD AND WATER USE

There are many examples of fertilization leading to increased WUE (e.g., Wong, Wild, and Mokwunye, 1991). Power, Grunes, and Reichmann (1961) reported that P fertilizer increased WUE of wheat, and Payne et al. (1995) found similar results for pearl millet. By increasing the slope of the line (yield as a function of evapotranspiration), fertilization increased the WUE. Nitrogen fertilization has been shown to increase WUE of native mixed prairie (Smika, Haas, and Power, 1965), wheat (Jensen and Sletten, 1965; Campbell et al., 1992), corn (Varvel, 1994), and sorghum (Onken et al., 1991; Varvel, 1995).

Yield

Fertilizer application normally results in increased yield with diminishing returns until maximum yield is reached; thereafter, excessive fertilizer application can reduce yield. Yield response to fertilization varies with crop, soil type, and other limiting factors. For example, in low-P soils, P fertilizer application increased millet dry matter yield and WUE (Payne et al., 1991). Nitrogen application to poinsettias increased dry matter production (Dole, Cole, and Von-Broembsen, 1994); there are many similar examples on a broad variety of plant species.

Use of legume cover crops in dryland cropping systems trades additional N supply for reduced water storage (Biederbeck and Bouman, 1994). However, in some cases, a legume may increase yield of the subsequent crop due to the additional N supply, resulting in improved WUE even though the legume utilized water which was, therefore, unavailable for the primary crop (Corak, Frye, and Smith, 1991). However, increasing WUE may also be limited by N supply. For example, in dryland crop rotation systems which increase yield and WUE, the need for supplemental N has been documented (Halvorson and Reule, 1994). In some cases, fertilization may increase dry matter production but have no effect on harvested yield. For example, applying K fertilizer to K-deficient peanuts can result in increased plant size and improved canopy cover without any subsequent effect on peanut yield (Davis, unpublished data). Therefore, increasing dry matter production alone does not increase WUE if harvested yield is unchanged. On the other hand, it is possible for fertilization to both increase yield and reduce HI in wheat (McNeal et al., 1971).

Yield quality can be affected by fertilization; for example, increased Mn application increases oil and decreases protein content in soybean seed and affects the oleic and linoleic acid contents as well (Wilson et al., 1982). Canola yields (and, subsequently, WUE) increase with N fertilization, but the oil content may decline (Taylor, Smith, and Wilson, 1991). Therefore, nutrients must be managed for optimum quality as well as quantity of yield, although yield quality is not reflected in WUE.

Nutrients can influence yield, and, therefore, WUE through their effect on photosynthesis. Nitrogen, S, and Mg are required for protein and chlorophyll synthesis in chloroplasts, and Mg, Fe, Cu, S, and P are essential in the electron transport chain (Marschner, 1986). In addition, some nutrients (Mg, Zn, Fe, K, Mn, P) have important roles in enzyme activation and osmoregulation in the photosynthetic process. Nitrogen deficiency has been shown to reduce foliar N levels, chlorophyll concentration, net photosynthesis, and WUE in sunflower (Fredeen, Gamon, and Field, 1991). The photosynthetic rate is diminished by N deficiency in jack pine (Tan and Hogan, 1995), and by both N and P deficiency in loblolly pine (Thomas, Lewis, and Strain, 1994).

The impact of nutrient stress on photosynthesis may increase susceptibility to an additional stress. For example, N deficiency has been shown to reduce photosynthesis of alfalfa under saline conditions (Khan, Silberbush, and Lips, 1994). On the other hand, a water stress may have more devastating consequences on photosynthesis and WUE under well-fertilized conditions than if nutrient availability is low (Kleiner, Abrams, and Schultz, 1992).

Plant nutrition can affect yield and, subsequently, WUE through its influence on flower initiation, flower fertilization, and seed development. For example, flower initiation in apples increases with ammonium application, due to the effect of ammonium on cytokinin transport from roots to shoots (Buban et al., 1978). Supply of P and K is also directly related to the cytokinin level and the flower number in many species (Marschner, 1986). Potassium deficiency can elevate abscisic acid levels in leaves, resulting in premature ripening and reduced seed size (Haeder and Beringer, 1981).

Nutritional impacts on plant reproduction influence yield and subsequently influence WUE. Nutritional effects on flower and seed development can be direct (deficiencies) or indirect, through the influence of a nutrient on a plant hormone. Flower fertilization can be severely reduced in cereal crops with Cu deficiency due to inhibition of anther formation, decreased number of pollen grains, and nonviability of the pollen (Graham, 1975). In addition, Mo deficiency decreases the number of pollen grains, and B is essential for pollen tube growth and silk receptiveness to pollen (Marschner, 1986).

In particular, N plays an important role in seed number and maturation. Nitrogen application prior to flower initiation has been found to increase the number of seeds per plant (Steer et al., 1984). Application of N fertilizer to soybeans during flowering has been shown to reduce pod drop and increase seed yield (Brevedan, Egli, and Leggett, 1978).

Evapotranspiration

Nutrients can influence ET of crops by altering (1) the available water supply and/or (2) the demand for water. Nutritional effects on root growth generally impact water availability by altering the potential for water uptake. Nutrient deficiency or toxicity symptoms which are expressed in the aboveground portion of the plant (for example, leaf area) influence the demand for water by the crop.

Supply of Water

Water supply can be increased (1) by improving rooting characteristics so that water present in the soil profile can be better utilized, and (2) by altering the soil water balance through indirect effects on infiltration rates and water loss pathways. Rooting volume and surface area are characteristics of root systems which can be manipulated to maximize available water supply (Taylor, 1983). Rooting volume is primarily a function of rooting depth and the rate of root extension. Not many cases of nutritional effects on the length of the main root axis have been reported. However, Ritchey, Silva and Costa (1982) determined that subsoil limitations to rooting in Brazilian Oxisols were related to Ca deficiency. The Ca/total cation ratio in soil solution was reported by Adams (1966) to be the determining factor in maximizing root length. Toxic levels of Mn, B, Zn, Cu, and Fe can also result in diminished root length (Davis, Hossner, and Persaud, 1993).

Rooting depth is an estimate of rooting volume at one point in time. The rate of root extension controls the length of time required to achieve the maximum rooting depth. Calcium deficiency can limit root extension rates. Calcium plays a critical role in cell extension and cell wall structure; therefore, root extension ceases in the absence of Ca (Marschner and Richter, 1974). In addition to Ca, deficiencies of P (Davis, Hossner, and Persaud, 1993) and B (Bohnsack and Albert, 1977) have also been reported to reduce root growth rates. On the other hand, toxic concentrations of ammoniacal N can reduce radicle elongation rates (Bennett and Adams, 1970).

Water use can be increased within an established rooting volume by increasing surface area of roots or rooting density. Water uptake is influenced by root length density to a much greater degree than by root weight, due to the close relationship between root length density and surface area of roots. Davis-Carter (1989) found that application of N (urea) and P (simple superphosphate) to sandy soils increased the root count (number of roots/surface area of soil pit face) of millet grown on those soils to a depth of 90 cm. Increased rooting density resulted in decreased soil water content due to increased water uptake. In this case, amelioration of a nutrient deficiency increased root density.

However, in other cases, nutrient deficiency has been reported to increase rooting. For example, Anghinoni and Barber (1980) reported that the longer the period of P starvation in maize, the greater the root length and the smaller the root radius. This adaptation results in greater surface area available for P absorption. Eghball and Maranville (1993) showed that the lack of N increased both root weight and length in maize.

Fertilizer placement can have a marked effect on rooting patterns as well. Sub-surface fertilizer banding has been shown to increase the root mass and, subsequently, the water use efficiency of barley and ryegrass (Sharratt, 1993; Murphy and Zaurov, 1994). However, water must be available in the banding zone to accomplish an increase in WUE.

The number and length of root hairs makes an important contribution to root surface area, which in turn influences water uptake. Soil nutrient status can affect root hair formation and growth. For example, Fe deficiency leads to abundant root hair formation and enhancement of Fe uptake (Römheld and Marschner, 1981). Phosphorus deficiency also results in increased number and length of root hairs (Föhse and Jungk, 1983). The form of N influences the relative abundance of root hairs. High nitrate concentrations result in reduction in number (Munns, 1968) and length (Bhat, Nye, and Brereton, 1979) of root hairs, but ammonium leads to the formation of abundant, long root hairs (Bhat, 1983).

Root function can also be affected by nutritional status. For example, P deficiency diminishes the hydraulic conductance of roots (Radin and Eidenbock, 1984). By reducing conductance, P deficiency effectively decreases the transport of water from soil through the roots to the leaves.

Fertilization can alter the amount of water available for plant uptake and evapotranspiration through its influence on the soil water balance. This is particularly true for organic fertilizer sources; by increasing soil OM content, these sources can increase water infiltration and hence enhance the availability of water in the root zone. Gypsum application can also increase infiltration by reducing soil crusting and runoff (Shainberg et al., 1989).

Nutrient supply can alter soil and water storage and availability through its effects on plant characteristics. For example, increased rooting depth due to proper fertilization could reduce deep drainage of water out of the root zone thus increasing water use by crops. On the other hand, nutritional enhancement of root extension could reduce storage of water in the soil for use later in the growing season; this could affect WUE adversely or positively depending on growth stage sensitivity to drought conditions.

Demand for Water

Nutritional impacts on water uptake by roots and transpiration from leaves can result in simultaneous, conflicting responses depending on other plant growth conditions. Nielsen and Halvorson (1991) reported that increasing N application rate increased the height, biomass, leaf area index, rooting depth, water use, and grain yield of winter wheat. Although N application increased rooting volume and reduced water stress when the temperature was low, when the temperature was high, higher N rates increased water stress due to greater leaf area and transpiration rates.

Nutritional status can affect transpiration rate during its influence on leaf area. Improvements in plant nutrition often lead to increased leaf area index. One example of this is the influence of N fertilizer on tillering of cereals; increased shoot density results in higher leaf area indices (Maizlish, Fritton, and Kendall, 1980). Eavis and Taylor (1979) reported that transpiration of soybeans increased with increasing leaf area. Ritchie and Burnett (1971) found that relative plant transpiration (T/ET) increased with leaf area index for both cotton and sorghum. As leaf area increases, transpiration increases and evaporation from the soil surface declines due to shading and canopy closure; however, this relationship is asymptotic and has reduced impact as leaf area increases. Therefore, nutrients influence transpiration demand through effects on leaf area index.

Leaf area duration influences transpiration differently throughout the growing season. Leaf growth and senescence rates influence the time required to reach the maximum leaf area and the time for that leaf area to decline. When the nutrient supply is deficient, the growth rate of the leaves can be limited by insufficient cell expansion or low photosynthetic rates (Marschner, 1986). Nitrogen and P deficiencies can both cause reduced cell expansion which results in smaller leaves. Application of N fertilizers can enhance new leaf growth (increased leaf area index) and delay plant senescence (increased leaf area duration), resulting in increased transpiration. In general, nutrient deficiencies lead to more rapid senescence.

Nutrient deficiency and toxicity symptoms are frequently reflected in leaf shape, angle, and color. For example, Mn toxicity results in leaf crinkling and cupping in soybeans. Although, the leaf area itself is not altered, the effective leaf area and hence the transpiration rate are reduced. Zinc toxicity in peanuts results in horizontal leaf orientation (Davis and Parker, 1993) which could result in increased absorption of radiant energy, increased leaf temperatures, and increased evaporation from leaf surfaces. On the other hand, nutritional inadequacies have been shown to cause bean leaflets to orient away from the sun's direct rays, thus reducing leaf conductance, transpiration, and WUE (Fu and Ehleringer, 1992).

Nutritional problems can also influence leaf color. Chlorosis is a symptom common to many nutritional deficiencies (e.g., N, Mo, and Fe). Phosphorus deficiency generally results in dark green leaves, and Mg deficiency can cause leaf reddening. These effects on leaf color may influence light absorption, leaf temperature, and photosynthetic rates. Nutrients may also influence leaf resistance to water loss through effects on the quantity and structure of hairs and cuticles on leaf surfaces (Marschner, 1986).

Nutrient effects on stomatal function can be direct, through the K^+ balance in guard cells, or indirect, through nutritional effects on abscisic acid production. Potassium ion is the major solute responsible for water potential gradients between guard cells of leaf stomata and surrounding epidermal cells (Zeiger, 1983). Potassium deficient plants have lower tolerance for water stress due to the role of K^+ in stomatal regulation and in plant cell vacuoles (Marschner, 1986). Under saline conditions, in particular, there is a high correlation between K^+ levels in plant shoots and stomatal conductance (Leidi et al., 1992). When Na, present in high concentration, competes with K for plant uptake, stomatal conductance is reduced; however, this may be an osmotic effect rather than a specific ion effect.

Improved plant nutrition increases both stomatal conductance and the assimilation rate (Radoglou, Aphalo, and Jarvis, 1992), but the ratio between these two factors determines the impact on WUE. For example, increased soil P levels increased transpirational WUE (photosynthetic rate/transpiration rate) of pearl millet whether or not water stress was present (Payne et al., 1992). However, in some cases improved N nutrition results in increased photosynthetic rates but does not effect stomatal control of water loss (Mitchell and Hinckley, 1993), thus resulting in higher WUE.

During periods of water stress, the abscisic acid level in leaves increases, and stomatal closure occurs, thus reducing transpiration. Goldbach et al. (1975) hypothesized that N deficiency enhanced abscisic acid synthesis and increased abscisic acid concentrations in leaves and stems. Radin and Ackerson (1981) reported that N deficiency resulted in more rapid stomatal closure and increased leaf resistance to water vapor diffusion, thus diminishing transpiration rates. Phosphorus-deficient plants also accumulate more abscisic acid in leaves than P-sufficient plants, and the stomata close at higher leaf water potentials, thus reducing transpiration (Radin, 1984). In addition, nutritional effects on leaf color, discussed previously, can also influence stomatal closure by altering light absorption and leaf temperature, and nutrition may also alter stomatal density and location (Marschner, 1986).

MANAGING NUTRIENTS TO TAKE ADVANTAGE OF CULTIVAR DIFFERENCES IN WUE

Considerable genetic variation has been recorded for WUE in many species (Quisenberry and McMichael, 1991; Donatelli, Hammer, and Vanderlip, 1992; Johnson and Tieszen, 1994). Cultivar tolerance to rooting limitations (such as compaction, Al toxicity, and salinity) increases indirectly the available water supply and optimizes the potential of a crop to respond to improved nutrition (Richards, 1993; Khan and Glenn, 1996). Both cultivar selection and crop fertilization practices can be chosen that will increase rooting volume or mass, and this increased rooting enhances the potential for a yield response due to additional water supply (Eghball and Maranville, 1993).

In addition, mycorrhizal inoculation of plant roots is known to increase effective root area, thus improving P status of the plant through increased P uptake. The ability of a crop to form mycorrhizal infection varies with cultivar, and increased infection enhances both water use efficiency and P status (Jun and Allen, 1991). Therefore, cultivar selection can influence both nutrient requirements and WUE.

Leaf traits play a critical role in water use patterns, and are similarly affected by both fertilization practices and cultivar selection. For example, leaf senescence (and, therefore, WUE) under drought conditions is known to vary with cultivar (Gwathmey and Hall, 1992). However, many nutritional deficiencies and toxicities can also stimulate early senescence (for example, N deficiency). Therefore, a cultivar with delayed senescence under drought conditions may also delay senescence under nutrient-deficient conditions, and this delayed senescence prolongs photosynthetic productivity and increases WUE.

In addition, specific leaf area is known to be positively correlated with water use efficiency and negatively correlated with leaf N concentration (Nageswara-Rao and Wright, 1994). Therefore, greater leaf area may increase productivity per unit land or unit water, but the nutrient content of the plant may be distributed throughout a larger plant and result in lower plant nutrient concentration.

Leaf color (and light absorption) also varies with both cultivar and nutrient status, and there may be potential for nutrients to be managed to make up for a less than optimum leaf color which is associated with a cultivar which has other desirable traits. For example, use of chlorophyll meters to determine N rates for corn has been complicated due to innate leaf color differences among cultivars; therefore, chlorophyll readings may have to be standardized across cultivars, although the N fertilizer recommendations may be the same (Waskom et al., 1996).

Grain characteristics also reveal potential for dual management of cultivar and fertilizer options. For example, kernel number is known to be related to preanthesis environmental stress. Entz and Fowler (1990) showed that wheat cultivar differences in water use efficiency were related to their ability to produce a high kernel number in spite of a preanthesis environmental stress (whether that be a water stress, nutritional stress, or pest-induced stress). Boron, K, and water deficiencies interfere with grain filling; therefore, although a cultivar may produce a greater kernel number, the kernels may not mature due to nutritional stresses.

Different genotypes achieve maximum grain yield at different fertilizer application rates. The water regime also alters the optimum fertilizer application rate. However, Eghball and Maranville (1991) found that genotypic selection of corn for nutrient use efficiency was not influenced by water regime, and that nutrient use efficiency closely paralleled WUE. This finding reveals the potential for genetic improvements in WUE through selection for nutrient use efficiency.

NEEDS FOR FURTHER RESEARCH

Increasing fertilizer use, in conjunction with improved varieties, has been responsible for growth in world food output since the 1940s (Brown, 1991). In the USA, the move toward sustainability and environmental stewardship has resulted in reduced fertilizer application rates; however, care must be taken not to jeopardize long-term sustainability for short-term benefits (Wallingford, 1991). A balanced approach with long-term and worldwide vision is the key to sustainability.

Some of the areas requiring additional research efforts include:

- Nutritional influences on transpiration
- Genetic variation in nutrient use efficiency and development of cultivars which will enhance nutrient uptake and WUE
- Relationship between nutrient use efficiency and tolerance to plant water stress (in particular, the role of root hairs and mycorrhizae)
- Fertilizer placement techniques to improve nutrient use efficiency and WUE
- Systems for altering nutrient management by cultivar, and
- Identification of cultivars which are tolerant of root limiting conditions.

It is critical that we meet the need for a systems approach in agricultural research (Logan, 1990) so that the issues listed above can be addressed successfully. Choosing the best fertilizer and cultivar combination can reduce the need for water, which in some situations (particularly, dryland crops), may be less dependable or available than fertilizer or seed. Therefore, it is essential to integrate the components of agricultural systems fully, so that their impact on other inputs is taken into account (Edwards, 1989). The more that scientists and farmers understand these interactions, the more capable we will be of making wise decisions regarding our stewardship of land and water.

REFERENCES

Adams, F. (1966). Calcium deficiency as a causal agent of ammonium phosphate injury to cotton seedlings. *Soil Science Society of America Proceedings* 30: 485-488.

Anghinoni, I. and S.A. Barber. (1980). Phosphorus influx and growth characteristics of corn roots as influenced by phosphorus supply. *Agronomy Journal* 72: 685-688.

Bennett, A.C. and F. Adams. (1970). Concentration of NH_3 (aq) required for incipient NH_3 toxicity to seedlings. *Soil Science Society of America Proceedings* 34: 259-263.

Bhat, K.K.S. (1983). Nutrient inflows into apple roots. *Plant and Soil* 71: 371-380.

Bhat, K.K.S., P.H. Nye, and A.J. Brereton. (1979). The possibility of predicting solute uptake and plant growth response from independently measured soil and plant characteristics. VI. The growth and uptake of rape in solutions of constant nitrate concentration. *Plant and Soil* 53: 137-167.

Biederbeck, V.O. and O.T. Bouman. (1994). Water use by annual green manure legumes in dryland cropping systems. *Agronomy Journal* 86: 543-549.

Blum, A. (1988). *Plant Breeding for Stress Environments*. Boca Raton, Florida, USA: CRC Press.

Bohnsack, C.W. and L.S. Albert. (1977). Early effects of boron deficiency on indoleacetic acid oxidase levels of squash root tips. *Plant Physiology* 59: 1047-1050.

Boyer, J.S. (1996). Advances in drought tolerance in plants. *Advances in Agronomy* 56: 187-218.

Brevedan, R.E., D.B. Egli, and J.E. Leggett. (1978). Influence of N nutrition on flower and pod abortion and yield of soybeans. *Agronomy Journal* 70: 81-84.

Brown, L.R. (1991). Feeding six billion. In *The Worldwatch Reader on Global Environmental Issues*, ed. L.R. Brown. New York, USA: Norton & Co., pp. 147-164.

Buban, T., A. Varga, J. Tromp, E. Knegt, and J. Bruinsma. (1978). Effects of ammonium and nitrate nutrition on the level of zeatin and amino nitrogen in xylem sap of apple rootstocks. *Zeitschrift für Pflanzenphysiologie* 89: 289-295.

Campbell, C.A., R.P. Zentner, B.G. McConkey, and F. Selles. (1992). Effect of nitrogen and snow management on efficiency of water use by spring wheat grown annually on zero-tillage. *Canadian Journal of Soil Science* 72: 271-279.

Corak, S.J., W.W. Frye, and M.S. Smith. (1991). Legume mulch and nitrogen fertilizer effects on soil water and corn production. *Soil Science Society of America Journal* 55: 1395-1400.

Davis, J.G., L.R. Hossner, and N. Persaud. (1993). Elemental toxicity effects on germination and growth of pearl millet seedlings. *Journal of Plant Nutrition* 16: 1957-1968.

Davis, J.G. and M.B. Parker. (1993). Zinc toxicity symptom development and partitioning of biomass and zinc in peanut plants. *Journal of Plant Nutrition* 16: 2353-2369.

Davis-Carter, J.G. (1989). Influence of spatial variability of soil physical and chemical properties on the rooting patterns of pearl millet and sorghum. Ph.D. Dissertation, Texas A&M University, College Station, Texas, USA.

Dole, J.M., J.C. Cole, and S.L. VonBroembsen. (1994). Growth of poinsettias, nutrient leaching, and water use efficiency respond to irrigation methods. *HortScience* 29: 858-864.

Donatelli, M., G.L. Hammer, and R.L. Vanderlip. (1992). Genotype and water limitation effects on phenology, growth, and transpiration efficiency in grain sorghum. *Crop Science* 32: 781-786.

Eavis, B.W. and H.M. Taylor. (1979). Transpiration of soybeans as related to leaf area, root length, and soil water content. *Agronomy Journal* 71: 441-445.

Edwards, C.A. (1989). The importance of integration in sustainable agricultural systems. *Agriculture, Ecosystems and Environment* 27: 25-35.

Eghball, B. and J.W. Maranville. (1991). Interactive effects of water and nitrogen stresses on nitrogen utilization efficiency, leaf water status and yield of corn genotypes. *Communications in Soil Science and Plant Analysis* 22: 1367-1382.

Eghball, B. and J.W. Maranville. (1993). Root development and nitrogen influx of corn genotypes grown under combined drought and nitrogen stresses. *Agronomy Journal* 85: 147-152.

Entz, M.H. and D.B. Fowler. (1990). Differential agronomic response of winter wheat cultivars to preanthesis environmental stress. *Crop Science* 30: 1119-1123.

Fischer, K.S., G.O. Edmeades, and E.C. Johnson. (1989). Selection for the improvement of maize yield under moisture deficits. *Field Crops Research* 22: 227-243.

Föhse, D. and A. Jungk. (1983). Influence of phosphate and nitrate supply on root hair formation of rape, spinach and tomato plants. *Plant and Soil* 74: 359-368.

Fredeen, A.L., J.A. Gamon, and C.B. Field. (1991). Responses of photosynthesis and carbohydrate-partitioning to limitations in nitrogen and water availability in field-grown sunflower. *Plant, Cell, and Environment* 14: 963-970.

Fu, Q.A. and J.R. Ehleringer. (1992). Paraheliotropic leaf movements in common bean under different soil nutrient levels. *Crop Science* 32: 1192-1196.

Goldbach, E., H. Goldbach, H. Wagner, and G. Michael. (1975). Influence of N deficiency on the abscisic acid content of sunflower plants. *Physiologia Plantarum* 34: 138-140.

Graham, R.D. (1975). Male sterility in wheat plants deficient in copper. *Nature* 254: 514-515.

Gwathmey, C.O. and A.E. Hall. (1992). Adaptation to midseason drought of cowpea genotypes with contrasting senescence traits. *Crop Science* 32: 773-778.

Haeder, H.E. and H. Beringer. (1981). Influence of potassium nutrition and water stress on the abscisic acid content in grains and flag leaves during grain development. *Journal of the Science of Food and Agriculture* 32: 552-556.

Haley, S.D. and J.S. Quick. (1993). Early generation selection for chemical desiccation tolerance in winter wheat. *Crop Science* 33: 1217-1223.

Halvorson, A.D. and C.A. Reule. (1994). Nitrogen fertilizer requirements in an annual dryland cropping system. *Agronomy Journal* 86: 315-318.

Jensen, M.E. and W.H. Sletten. (1965). Evapotranspiration and soil moisture-fertilizer interrelations with irrigated winter wheat in the southern High Plains. USDA/TAES Conserv. Res. Rep. no. 4.

Johnson, R.C. and L.L. Tieszen. (1994). Variation for water-use efficiency in alfalfa germplasm. *Crop Science* 34: 452-458.

Jun, D.J. and E.B. Allen. (1991). Physiological responses of 6 wheatgrass cultivars to mycorrhizae. *Journal of Range Management* 44: 336-340.

Khan, M.G., M. Silberbush, and S.H. Lips. (1994). Physiological studies on salinity and nitrogen interaction in alfalfa. II. Photosynthesis and transpiration. *Journal of Plant Nutrition* 17: 669-682.

Khan, M.J. and E.P. Glenn. (1996). Yield and evapotranspiration of two barley varieties as affected by sodium chloride salinity and leaching fraction in lysimeter tanks. *Communications in Soil Science and Plant Analysis* 27: 157-177.

Kleiner, K.W., M.D. Abrams, and J.C. Schultz. (1992). The impact of water and nutrient deficiencies on the growth, gas exchange and water relations of red oak and chestnut oak. *Tree Physiology* 11: 271-287.

Lande, R. and R. Thompson. (1990). Efficiency of marker-assisted selection in the improvement of quantitative traits. *Genetics* 124: 743-756.

Leidi, E.O., M. Silberbush, M.I.M. Soares, and S.H. Lips. (1992). Salinity and nitrogen nutrition studies on peanut and cotton plants. *Journal of Plant Nutrition* 15: 591-604.

Logan, T.J. (1990). Sustainable agriculture and water quality. In *Sustainable Agricul-

tural Systems, ed. C.A. Edwards. Ankeny, Iowa, USA: Soil and Water Conservation Society, pp. 582-613.

Maizlish, N.A., D.D. Fritton, and W.A. Kendall. (1980). Root morphology and early development of maize at varying levels of nitrogen. *Agronomy Journal* 72: 25-31.

Marschner, H. (1986). *Mineral Nutrition of Higher Plants*. London, UK: Academic Press.

Marschner, H. and C. Richter. (1974). Calcium-transport in wurzeln von mais-und bohnenkeimpflanzen. *Plant and Soil* 40: 193-210.

Martin, B., J. Nienhuis, G. King, and A. Shaefer. (1989). Restriction fragment length polymorphisms associated with water use efficiency in tomato. *Science* 243: 1725-1728.

McNeal, F.H., M.A. Berg, P.L. Brown, and C.F. McGuire. (1971). Productivity and quality response of five spring wheat genotypes, *Triticum aestivum* L., to nitrogen fertilizer. *Agronomy Journal* 63: 908-910.

Mian, M.A.R., M.A. Bailey, D.A. Ashley, R. Wells, T.E. Carter Jr., W.A. Parrott, and H.R. Boerma. (1996). Molecular markers associated with water use efficiency and leaf ash in soybean. *Crop Science* 36: 1252-1257.

Mitchell, A.K. and T.M. Hinckley. (1993). Effects of foliar nitrogen concentration on photosynthesis and water use efficiency in Douglas fir. *Tree Physiology* 12: 403-410.

Morgan, J.A., G. Zerbi, M. Martin, M.Y. Mujahid, and J.S. Quick. (1993). Carbon isotope discrimination and productivity in winter wheat. *Journal of Agronomy and Crop Science* 171: 289-297.

Morgan, J.M. (1983). Osmoregulation as a selection criterion for drought tolerance in wheat. *Australian Journal of Agricultural Research* 34: 607-614.

Munns, D.N. (1968). Nodulation of *Medicago sativa* in solution culture. III. Effects of nitrate on root hairs and infection. *Plant and Soil* 29: 33-47.

Murphy, J.A. and D.E. Zaurov. (1994). Shoot and root growth response of perennial ryegrass to fertilizer placement depth. *Agronomy Journal* 86: 828-832.

Nageswara-Rao, R.C. and G.C. Wright. (1994). Stability of the relationship between specific leaf area and carbon isotope discrimination across environments in peanut. *Crop Science* 34: 98-103.

Nielsen, D.C. and A.D. Halvorson. (1991). Nitrogen fertility influence on water stress and yield of winter wheat. *Agronomy Journal* 83: 1065-1070.

Onken, A.B., C.W. Wendt, R.J. Lascano, A. Sow, and Z. Kouyate. (1991). Effects of soil fertility, crop genotypes, and available soil water on water-use efficiency of sorghum. In *TropSoils Technical Report, 1988-1989*. Raleigh, North Carolina, USA: North Carolina State University, pp. 318-320.

Payne, W.A., M.C. Drew, L.R. Hossner, R.J. Lascano, A.B. Onken, and C.W. Wendt. (1992). Soil phosphorus availability and pearl millet water use efficiency. *Crop Science* 32: 1010-1015.

Payne, W.A., L.R. Hossner, A.B. Onken, and C.W. Wendt. (1995). Nitrogen and phosphorus uptake in pearl millet and its relation to nutrient and transpiration efficiency. *Agronomy Journal* 87: 425-431.

Payne, W.A., A.B. Onken, C.W. Wendt, and R.J. Lascano. (1991). Effects of soil phosphorus and water supply on growth analysis and transpirational water-use efficiency of pearl millet. In *TropSoils Technical Report, 1988-1989.* Raleigh, North Carolina, USA: North Carolina State University, pp. 320-322.

Power, J.F., D.L. Grunes, and G.A. Reichmann. (1961). The influence of phosphorus fertilization and moisture on growth and nutrient absorption by spring wheat: I. Plant growth, N uptake, and moisture use. *Soil Science Society of America Proceedings* 25: 207-210.

Quisenberry, J.E. and B.L. McMichael. (1991). Genetic variation among cotton germplasm for water-use efficiency. *Environmental and Experimental Botany* 31: 453-460.

Radin, J.W. (1984). Stomatal responses to water stress and to abscisic acid in phosphorus-deficient cotton plants. *Plant Physiology* 76: 392-394.

Radin, J.W. and R.C. Ackerson. (1981). Water relations of cotton plants under nitrogen deficiency. III. Stomatal conductance. *Plant Physiology* 67: 115-119.

Radin, J.W. and M.P. Eidenbock. (1984). Hydraulic conductance as a factor limiting leaf expansion of phosphorus-deficient cotton plants. *Plant Physiology* 75: 372-377.

Radoglou, K.M., P. Aphalo, and P.G. Jarvis. (1992). Response of photosynthesis, stomatal conductance and water use efficiency to elevated CO_2 and nutrient supply in acclimated seedlings of *Phaseolus vulgaris* L. *Annals of Botany* 70: 257-264.

Richards, R.A. (1987). Physiology and the breeding of winter-grown cereals for dry areas. In *Drought Tolerance in Winter Cereals*, ed. J.P. Srivastava et al. London, UK: Wiley, pp. 133-150.

Richards, R.A. (1993). Increasing salinity tolerance of grain crops: is it worthwhile? *Developments in Plant and Soil Science* 50: 117-126.

Richards, R.A. and J.B. Passioura. (1981). Seminal root morphology and water use in wheat. *Crop Science* 21: 253-255.

Ritchey, K.D., J.E. Silva, and U.F. Costa. (1982). Calcium deficiency in clayey B horizon of savanna Oxisols. *Soil Science* 133: 378-382.

Ritchie, J.T. and E. Burnett. (1971). Dryland evaporative flux in a subhumid climate. II. Plant influences. *Agronomy Journal* 63: 56-62.

Römheld, V. and H. Marschner. (1981). Rhythmic iron stress reactions in sunflower at suboptimal iron supply. *Physiologia Plantarum* 53: 347-353.

Schneider, K.A., M.E. Brothers, and J.D. Kelly. (1997). Marker-assisted selection to improve drought resistance in common bean. *Crop Science* 37: 51-60.

Shainberg, I., M.E. Sumner, W.P. Miller, M.P.W. Farina, M.A. Pavan, and M.V. Fey. (1989). Use of gypsum on soils: A review. *Advances in Soil Science* 9: 1-111.

Sharratt, B.S. (1993). Water use, intercepted radiation, and soil temperature of skip-row and equidistant-row barley. *Agronomy Journal* 85: 686-691.

Smika, D.E., H.J. Haas, and J.F. Power. (1965). Effects of moisture and nitrogen fertilizer on growth and water use by native grass. *Agronomy Journal* 57: 483-486.

Steer, B.T., P.J. Hocking, A.A. Kortt, and C.M. Roxburgh. (1984). Nitrogen nutrition of sunflower (*Helianthus annus* L.) yield components, the timing of their establishment and seed characteristics in response to nitrogen supply. *Field Crops Research* 9: 219-236.

Tan, W. and G.D. Hogan. (1995). Limitations to net photosynthesis as affected by nitrogen status in jack pine (*Pinus banksiana* Lamb.) seedlings. *Journal of Experimental Botany* 46: 407-413.

Taylor, A.J., C.J. Smith, and I.B. Wilson. (1991). Effect of irrigation and nitrogen fertilizer on yield, oil content, nitrogen accumulation and water use of canola (*Brassica napus* L.). *Journal of Fertilizer Use Technology* 29: 249-260.

Taylor, H.M. (1983). Managing root systems for efficient water use: an overview. In *Limitations to Efficient Water Use in Crop Production*, eds. H.M. Taylor, W.R. Jordan, and T.R. Sinclair. Madison, Wisconsin, USA: American Society of Agronomy, pp. 87-113.

Thomas, R.B., J.D. Lewis, and B.R. Strain. (1994). Effects of leaf nutrient status on photosynthetic capacity in loblolly pine (*Pinus taeda* L.) seedlings grown in elevated atmospheric CO_2. *Tree Physiology* 14: 947-960.

Turner, N.C., M.E. Nicolas, K.T. Hubick, and G.D. Farquhar. (1989). Evaluation of traits for the improvement of water use efficiency and harvest index. In *Drought Resistance in Cereals*, ed. F.W.G. Baker. Wallingford, UK: CAB International, pp. 177-189.

Varvel, G.E. (1994). Monoculture and rotation system effects on precipitation use efficiency of corn. *Agronomy Journal* 86: 204-208.

Varvel, G.E. (1995). Precipiation use efficiency of soybean and grain sorghum in monoculture and rotation. *Soil Science Society of America Journal* 59: 527-531.

Viets, F.G., Jr. (1962). Fertilizers and the efficient use of water. *Advances in Agronomy* 14: 223-264.

Wallingford, G.W. (1991). The U.S. nutrient budget is in the red. *Better Crops with Plant Food* 75: 16-18.

Waskom, R.M., D.G. Westfall, D.E. Spellman, and P.N. Soltanpour. (1996). Monitoring nitrogen status of corn with a portable chlorophyll meter. *Communications in Soil Science and Plant Analysis* 27: 545-560.

Wilson, D.O., F.C. Boswell, K. Ohki, M.B. Parker, L.M. Shuman, and M.D. Jellum. (1982). Changes in soybean seed oil and protein as influenced by manganese nutrition. *Crop Science* 22: 948-952.

Wong, M.T.F., A. Wild, and A.U. Mokwunye. (1991). Overcoming soil nutrient constraints to crop production in West Africa: importance of fertilizers and priorities in soil fertility research. *Fertilizer Research* 29: 45-54.

Zeiger, E. (1983). The biology of stomatal guard cells. *Annual Review of Plant Physiology* 34: 441-475.

SUBMITTED: 01/14/97
ACCEPTED: 06/27/97

The Role of Nutrient-Efficient Crops in Modern Agriculture

Jonathan Lynch

SUMMARY. Nutrient-efficient crops have an important role in modern agriculture. In the low-input systems that characterize most of world agriculture, nutrient-efficient crops improve crop productivity. In high-input systems of the developed world, nutrient-efficient crops are valuable in reducing pollution of surface and ground water resources from intense fertilization. Recent developments in molecular biology, root biology, rhizosphere interactions, and modeling present new opportunities for the understanding and improvement of crop nutrient efficiency. The degree and extent of nutritional limitations to crop productivity, and the economic and ecological liabilities of intensive fertilization, are such that eventually nutrient-efficient crops will be an important part of integrated nutrient management of cropping systems. *[Article copies available for a fee from The Haworth Document Delivery Service: 1-800-342-9678. E-mail address: getinfo@haworthpressinc.com]*

KEYWORDS. Genotype, genotypic differences, germplasm, nutrient efficiency, plant nutrition

INTRODUCTION

The purpose of this chapter is to consider some of the issues surrounding the potential role and value of nutrient-efficient crops in

Jonathan Lynch, Associate Professor of Plant Nutrition, Department of Horticulture, Pennsylvania State University, University Park, PA 16802 USA (E-mail: jlynch@psupen.psu.edu).

Financial support was provided by USDA/NRI grants 94371000311 and 9700533.

[Haworth co-indexing entry note]: "The Role of Nutrient-Efficient Crops in Modern Agriculture." Lynch, Jonathan. Co-published simultaneously in *Journal of Crop Production* (Food Products Press, an imprint of The Haworth Press, Inc.) Vol. 1, No. 2 (#2), 1998, pp. 241-264; and: *Nutrient Use in Crop Production* (ed: Zdenko Rengel) Food Products Press, an imprint of The Haworth Press, Inc., 1998, pp. 241-264. Single or multiple copies of this article are available for a fee from The Haworth Document Delivery Service [1-800-342-9678, 9:00 a.m. - 5:00 p.m. (EST). E-mail address: getinfo@haworthpressinc. com].

modern agriculture. I will not attempt a comprehensive treatment of agronomic, physiological, or genetic aspects of crop nutrient efficiency. The review by Clark (1990) remains an excellent overall treatment of nutrient efficiency in cereals. As examples of micronutrient efficiency, Fe efficiency in crops has been reviewed by Brown and Jolley (1989), and Zn efficiency by Graham, Ascher, and Hynes (1992). As examples of macronutrient efficiency, the case of P efficiency in bean has been recently reviewed by Lynch and Beebe (1995), and P efficiency in wheat by Batten (1992), while the utility of N efficiency in tropical maize has been treated by Feil, Thiraporn, and Stamp (1992). These and other similar reviews confirm that our understanding of the agronomic, physiological, and genetic bases of crop nutrient efficiency is growing. The potential impact of these advances is largely unrealized, however. Relatively little effort is devoted to the improvement of crop nutrient efficiency despite growing recognition of the importance of integrated nutrient management in both high-input and low-input systems. This is in large part because many policymakers, funding agencies, and agricultural scientists either are not aware of nutrient efficiency in crops or doubt its value. In this article I hope to address this lack of awareness and counter some of the more common doubts about the value of nutrient-efficient crops in modern agriculture.

What Is Nutrient Efficiency?

Nutrient efficiency has been defined in many ways in diverse contexts (Clark, 1990; Blair, 1993). Most definitions share the concept that efficiency is the ability of a system to convert inputs into desired outputs, or to minimize the conversion of inputs into waste (Figure 1). When considering nutrient efficiency, supply or availability or amount of a mineral nutrient is the input, and plant growth, physiological activity, or yield are typical outputs. "Efficiency" is the relationship of output to input. In some cases this is expressed as a simple ratio, such as kg yield per kg fertilizer, or g of plant dry weight per mg of nutrient (the inverse of nutrient concentration, as discussed in Clark, 1990), but as the amount of input and output vary, the ratio between them is rarely fixed, so efficiency is most comprehensively described by the entire relationship of output as a function of input (e.g., Figure 2).

This definition of efficiency has the greatest validity when the system in question has organic integrity (i.e., a discrete unit with closely

FIGURE 1. Diagram illustrating a conceptual approach to nutrient efficiency, defined as the ability of a system to convert nutrient inputs into desired outputs, thereby minimizing nutrient waste.

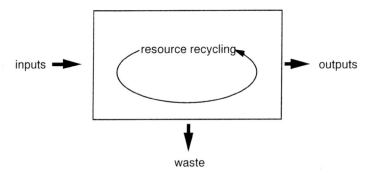

FIGURE 2. Strategies for improved nutrient efficiency relative to a standard response curve (thick line): (1) the green revolution, or improving yield response to high nutrient inputs, (2) improving yield response to low nutrient availability, and (3) improving yield response to both low and high nutrient inputs.

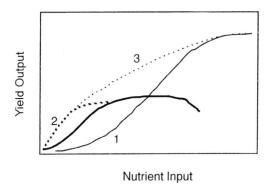

interrelated processes), and when the input has a direct relationship with the output of interest. For example, the ability of a leaf to utilize N for photosynthesis is a useful index of nutrient efficiency since a leaf is an obvious functional unit and N content is a primary determinant of photosynthetic capacity (e.g., Lynch and Rodriguez, 1994). The ability of a leaf to utilize Ca for photosynthesis is less meaningful since leaf Ca content is less directly related to photosynthetic capacity (e.g., see discussion in Lynch and González, 1993). The establishment of causal links between an input and an output can be challenging in complex systems. The principle complication in studies of this sort is the soil itself, which has many complex and variable abilities to sequester and release applied nutrients that may obscure efficiency characteristics of the crop. For example, crops grown in P-fixing soils may be inefficient at converting applied P fertilizer into yield, simply because most of the applied fertilizer is unavailable to the crop (Sample, Soper, and Racz, 1980), regardless of the inherent ability of the crop to acquire and utilize available P in producing yield. Additional complications arise when factors other than nutrient availability limit yield, thereby uncoupling nutrient processes from growth. A clear and logical delineation of a functional system, and functional links between an input and an output within that system, are important yet often overlooked elements in understanding crop nutrient efficiency.

Since nutrient efficiency refers to the ability of a plant system to convert nutrient inputs into desired outputs, every plant has a 'nutrient efficiency,' just as every automobile has a 'fuel efficiency.' The term nutrient efficiency is most commonly used, however, as a comparative term to distinguish plant systems that differ in their ability to convert nutrient inputs into desired outputs. In this context a 'nutrient-efficient' genotype is one that better converts nutrient inputs into desired outputs than other genotypes, which by comparison are 'nutrient-inefficient.' For the sake of simplicity I will continue to use a plant genotype as an example system for the discussion of nutrient efficiency; similar considerations would apply to the analysis of nutrient efficiency in plant organs, physiological processes, crops, cropping systems, etc.

Figure 2 illustrates three ways in which a genotype may have superior nutrient efficiency. When nutrient efficiency is defined as the function of desired output over nutrient input, an efficient genotype is one that has greater output over some significant range of nutrient input. An important historical example of this is shown by line 1 in Figure 2, in which the response of genotypes to high levels of nutrient input was improved. This was the basis for the 'green revolution,' in which short-statured high-yielding varieties of rice and wheat were developed that could respond to high rates of N fertilization without lodging (Borlaug, 1972). In contrast, traditional cultivars were adapted to low input conditions, and would lodge when excessively fertilized, thereby reducing yield. The role and value of nutrient-efficient crops in modern agriculture is clearly shown by the fact that perhaps the single most dramatic breakthrough in world agriculture this century, the green revolution, was basically an improvement in the nutrient efficiency of wheat and rice.

An often-criticized aspect of the 'green revolution' varieties is that the new varieties were less efficient at low levels of nutrients, as shown in Figure 2. In this way green revolution technology was 'scale-positive' in economic parlance, in that it was a technology that disproportionately benefited larger or more wealthy producers that could afford fertilizers. This accentuated resource inequities among agricultural producers and drove many small farmers off their land (Shiva, 1991). An alternative approach to enhancing crop nutrient efficiency is shown by the second curve in Figure 2. In this case the yield response to nutrients over the low range of nutrient availability is increased, without affecting the response to high rates of nutrient input. This type of nutrient efficiency is obviously more useful in situations of low nutrient availability, as in low-input agriculture. The common bean is a classic example of a crop that would benefit from this type of efficiency, because it is a subsistence crop that is often grown in marginal lands with few inputs (Schwartz and Pastor-Corrales, 1989). The types of nutrient efficiency shown in curves 1 and 2 of Figure 2 are sometimes given distinct terms: the ability to grow or yield at low nutrient availability has been called 'efficiency,' while the capacity to respond to increasing levels of nutrient availability has been called 'responsiveness.' This was the terminology used in the classic studies of Gerloff and Gabelman in their studies of P efficiency in common bean (Whiteaker et al., 1976). This is a useful functional distinction since constitutive traits conferring adaptation to low nutrient availability, such as a high growth allocation to root production, may be counterproductive at high nutrient availability. Contrasting responses to low or high levels of nutrient availability may also be related to differences in the physiological mechanism conferring overall nutrient efficiency in a genotype. Traits improving nutrient acquisition may confer efficiency at low levels of nutrient availability, while traits improving the internal utilization of acquired nutrients may be useful at all levels of nutrient availability. A special case of curve 2 occurs if greater response to low levels of nutrients is achieved in genotypes with high yield ability. Such genotypes may be particularly useful in high input systems, as discussed below.

The third type of nutrient efficiency is superior yield response at all levels of nutrient availability, shown by curve 3 in Figure 2. Genotypes showing this response are highly valued in selection and breeding programs. Broad adaptation across diverse environments may be due to factors that are not directly related to nutrient acquisition or utilization, such as better photoperiod/temperature adaptation, better shoot architecture, more appropriate phenology, indeterminacy, broad disease resistance, Al tolerance, etc. Curve 3 could also result from traits improving the utilization of acquired nutrients, or from traits permitting the acquisition of nutrients with reduced expenditure of plant resources, such as might be conferred by changes in root architecture or morphology (Fitter, 1991; Fitter et al., 1991; Lynch, 1995).

In actual practice few researchers have the opportunity to study the response of a set of genotypes at many different levels of nutrient availability, especially in field studies. In most cases just two or three rates of fertilizer are employed, and the yield compared at a high versus a low level of nutrient availability (Figure 3; see also Graham, 1984). The resulting figure can be divided into four quadrants for the identification of genotypes that yield well at both high and low nutrient availability (Thung, 1990).

The potential importance of nutrient efficiency is dependent on the level of nutrient inputs provided to the system. Therefore, nutrient efficiency has different implications for cropping systems in different sectors of modern agriculture, as discussed next.

NUTRIENT-EFFICIENT CROPS
IN MODERN AGRICULTURE

The term 'modern agriculture' may call to mind images of high-technology production systems that can be found in parts of the wealthier nations of the world. This is 'modern' in the sense of 'up-to-date.' I choose the broader meaning of the word 'modern,' however, as a synonym of 'contemporary.' 'Modern agriculture' in this sense is agriculture in late 90's, which includes high input as well as decidedly lower input production systems. In the present discussion I will follow the analytical convention adopted by the World Bank (World Bank, 1996) in dividing the nations of the world into three categories; high income economies ($8,956 or more GNP per capita in 1994), middle income economies (between $726 and $8,955 GNP per capita in

FIGURE 3. Example of identification of genotypes as efficient or responsive based on relative yield at high and low levels of nutrient supply (each point represents the mean value for a genotype). The Figure also illustrates the positive correlation of yield at high and low nutrient supply commonly observed in field studies. This correlation may be due to general adaptation to the selection environment rather than specific traits for nutrient efficiency, illustrating the potential utility of trait-based selection.

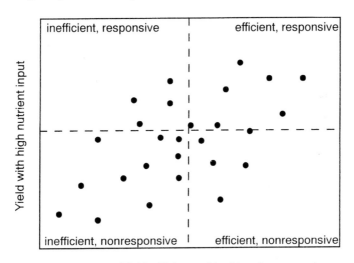

Yield with low nutrient input

1994), and low income economies (less than $726 GNP per capita in 1994). Low and middle income economies are also termed 'developing countries' or 'Less Developed Countries.' Economic and demographic data cited below are for 1994 (World Bank, 1996).

High Income Economies

This group includes the United States, Canada, Western Europe, Australia, New Zealand, Japan, Israel, and various islands, territories, and oil states. This group of nations represents 24% of the earth's land area, and approximately 15% of the global population (850×10^6 out of 5.6×10^9 people) producing an average GNP per capita of $23,420. Only 5% the workforce is employed in agriculture, which accounts for less than 3% of the economies of these nations.

Within high income economies (as also in Less Developed Countries), there is considerable diversity of production systems with respect to nutrient utilization. In general, however, these systems are characterized by high nutrient availability, both because of the prevalence of fertile soils in North America and Europe, as well as decades of nutrient application in the form of fertilizers or manures. In some sectors (such as intensive horticultural production) nutrient application is very high, but even in these cases the cost of nutrient amendments are typically a small proportion of total production costs.

Nutrient efficiency in these systems would probably not improve yields, which are limited by other, less controllable factors such as weather. Genotypes employed in high input agriculture are typically 'responsive' to fertilizer yet 'inefficient' at low input rates, as shown in curve 1 of Figure 2. In this case the improvement of nutrient efficiency in the manner of curve 3 in Figure 2 would permit the reduction of fertilizer rates with maintenance of high yields. This would have economic benefits for producers by reducing fertilizer costs. Although savings realized in this way may be a minor component of overall production costs, in systems with low profit margins this savings may be consequential. In some systems, such as extensive pasture grazing in low-P soils of New Zealand, profit margins are too thin to justify adequate fertilization rates. In such cases nutrient-efficient crops could improve profitability, and are the target of crop breeding programs. Nutrient efficiency in the developed world has much the same role in integrated nutrient management that cultivar disease resistance has in IPM programs, i.e., one component of an overall management system (Callaway and Francis, 1993).

Additional benefits from reduced fertilization may result from improved crop quality. There is much evidence that plants given abundant nutrients may be more susceptible to diseases, insects, and abiotic stress such as drought than plants given merely adequate nutrition (Marschner, 1995). Indeed, it is one of the tenants of 'organic agriculture' that crops produced with high rates of fertilization are less healthy and less nutritious than plants produced more naturally. While a full discussion of the technical merits of this claim is beyond the scope of this review, it is sufficient to note that crops produced 'organically' are one of the fastest growing segments of US agriculture (anonymous, 1996) show that the claim has social and economic validity.

The most important role that nutrient-efficient crops could play in wealthy countries is not in the improvement of crop yield, quality, or profitability. The greatest potential impact of nutrient-efficient crops in wealthy nations is rather in the reduction of nutrient wastes (Figure 1). Excessive N fertilization has led to nitrate contamination of groundwater resources in large areas of North America and Western Europe (Hallberg, 1989; Jürgens-Gschwind, 1989). Nitrate in groundwater is a human health concern because it is linked with stomach cancer and is harmful to infants (Fraser and Chilvers, 1981; CAST, 1985; Follett and Walker, 1989). Excessive P fertilization is linked with runoff contamination of surface water resources, where it stimulates algal blooms (eutrophication) with devastating ecological consequences (Ryden, Syers, and Harris, 1973). An interesting development in this regard is the discovery that effluent from the Mississippi River Basin into the Gulf of Mexico during the 1993 summer storms has produced large, persistent eutrophic areas of the sea floor (Louisiana State University, 1995). In areas with intense animal production, such as the Northeastern US and parts of Europe, nutrient overloading from animal wastes is particularly important, since intensive animal production concentrates nutrients in the form of feed materials in small areas (NRC, 1989). Countries such as Denmark and the Netherlands have strict nutrient management programs that must be followed by agricultural producers in order to minimize overloading of the soil and water resources with excess N and P, and similar programs have been contemplated by regulatory agencies in the US.

Nutrient-efficient crops, especially those acting through aggressive nutrient acquisition rather than more efficient nutrient utilization in the plant, could help alleviate these problems by reducing nonpoint sources of pollutants in the regions of crop production, and by improved N and P scavenging in regions of manure application. More efficient nutrient scavenging might be realized through agronomic means, such as intercropping and cover cropping so that root activity through the year was maximized. Crop genotypes with more vigorous or prolonged nutrient uptake might also be useful. This would be particularly useful in the case of N scavenging in extensive agronomic crops such as maize and wheat.

Middle Income Economies

This group includes some of Southern Africa, Northern Africa, the Middle East (excluding Israel and the oil states), most of Eastern Europe, and most of Latin America. This group of nations represents 46% of the earth's land area, and approximately 28% of the global population (1.6×10^9 out of 5.6×10^9 people) producing an average GNP per capita of $2,520. About 31% the workforce is employed in agriculture, which accounts for 10% of the economies of these nations.

Agricultural production in this region is often highly diverse, with some sectors, especially export-oriented high-value crops, receiving high nutrient inputs, and other sectors, notably extensive systems and subsistence farming, receiving few nutrient inputs. In the arid and semi-arid regions of this group, including Northern Africa and the Middle East, water availability is the principal production constraint. In such areas nutrient-efficient crops probably would have little potential impact. In many parts of Eastern Europe fertilizer use has declined sharply with the economic dislocations and loss of agricultural subsidies in recent years (FAO, 1996a, b). If nutrient-efficient crops were available immediately they may be useful in maintaining yields with declining inputs. These nations are in a state of rapid socioeconomic change, however, and by the time nutrient-efficient crops could be developed for them, their agricultural sectors may have problems more characteristic of the wealthy nations of Western Europe. This is particularly true as regards environmental contamination with fertilizer runoff.

Large parts of Latin America have weathered soils with low inherent fertility and poor responsiveness to applied fertilizer, especially to P (Uehara and Keng, 1975). Latin America also has great disparity among agricultural producers, from high-technology export-oriented horticultural production to traditional subsistence cultivation of staple crops. In many Latin American countries, subsistence farmers receive little development aid, while high-technology producers receive little regulatory oversight. Nutrient-efficient crops would have potential value in both types of systems, by improving productivity in nutrient-limited systems and by reducing environmental contamination in nutrient-rich systems. Nutrient costs are often a greater proportion of total production costs in Latin America than in high-income economies, so nutrient-efficient crops could also benefit high-input systems by increasing profitability. Although many traditional subsistence systems are already efficient in acquiring and utilizing scarce nutrient resources through such means as intercropping, crop selection, and residue management, population pressures have degraded soil fertility and precluded many traditional fertility management options. In such systems the introduction of more nutrient-efficient genotypes or more nutrient-efficient species would boost productivity. For example, cassava (*Manihot esculenta* L.) yields well on soils that would not support more traditional crops such as maize and bean, and is suited for cultivation of degraded or marginal lands (Cock, 1985). Aluminum-tolerant genotypes of wheat, maize, and soybean are converting what were once considered marginal lands into productive areas in the vast areas of Latin America in savannahs (the llanos) and scrub lands (the cerrado of Brazil) (Kohli and Rajaram, 1988). Nutrient efficiency is one aspect of the adaptation of these crops to the acid soils and minimal fertilization regimes employed in these agricultural frontiers. Development of such lands for crop production may be helpful in slowing the rate of clearing of fragile and ecologically valued tropical forests (Hecht, 1982).

Low Income Economies

This group includes most of sub-Saharan Africa, Egypt, most of East Asia including China, most of South Asia including India, and parts of Eastern Europe, Central Asia, and Central America. These nations represents about 30% of the earth's land area, yet approximately 60% of the global population (3.2×10^9 out of 5.6×10^9 people), producing an average GNP per capita of $380. Roughly 70% of the workforce is employed in agriculture, which accounts for 28% of the economies of these nations.

In many areas such as South China and sub-Saharan Africa, native soil fertility is low, and population pressures have degraded soil fertility through erosion and fertility depletion (Sanchez and Buol, 1975; Sanchez, 1976). Nutrient inputs are generally low, or if used may not be optimally suited to soil conditions. Extension of agricultural technology into rural areas ranges from good, as in China, to ineffectual, as in many African countries.

In less developed countries, nutrient-efficient crops would have substantial impact in improving crop yields under nutrient limited conditions (i.e., curve 2 in Figure 2). Current average yields for crops grown in this region are often as low as 10 or 20% of the potential yield with high inputs; much of the lost yield is from drought and nutrient stress. Although accurate estimates of the extent of yield loss from nutrient limitation are scarce, careful GIS studies with common bean showed that at least 90% of bean production in Africa is limited by nutrient deficiency (Wortmann and Allen, 1994). Nutrient-efficient genotypes would represent 'appropriate technology' in these systems since they would improve productivity without requiring increased inputs of information, labor, or material resources; could be easily disseminated by resource poor producers with existing infrastructure; and unlike pest-resistant genotypes, would not require continued development assistance to remain effective.

A frequent objection to the introduction of nutrient-efficient crops in low-input systems is that such crops will 'mine the soil,' thereby degrading system productivity in the long run. Nutrient efficiency based on more efficient nutrient utilization rather than more aggressive nutrient extraction would not be subject to this concern. It is likely that nutrient efficiency through more aggressive nutrient acquisition would be easier to obtain in most cases, because of the inherent efficiencies of plant metabolism (Gutschick, 1993). Although much evidence shows that continued crop cultivation without replenishments of nutrients may indeed deplete soil nutrients, as far as I am aware there are no studies comparing the relative impacts of nutrient-efficient and nutrient-inefficient crops on nutrient cycling processes.

In agronomic crops, the grain may be exported from the farm, and other crop residues are usually returned to the soil or used by livestock on farm. When residues are burned, N and S are lost but not other nutrients. Legumes tend to have the greatest seed mineral content. Legumes would typically replace exported seed N through fixation. The relatively small amount of P and K exported in grain (compared to the amount of available P and K in the system) may well be offset by the soil fertility benefits from greater crop growth. On slopes, the greatest such benefit could be reduced soil erosion, which in the humid tropics can represent the loss of hundreds of tonnes of topsoil per hectare each year, with accompanying loss of hundreds of kg of nutrients (Lal, 1985). Greater crop biomass would also lead to greater production of soil organic matter, greater scavenging of mobile nutrients that may otherwise be lost to leaching, less nutrient competition from weeds, greater availability of mulch or green manure from crop residues, and for legumes, greater N fixation. The net effect of greater crop nutrient extraction, and hence greater crop biomass production, on soil fertility may thus be positive. In nonleguminous crops, extra N extraction would have to be offset by the use of legume rotations, and in some high biomass crops such as plantain and cassava, prolonged K extraction can be problematic. Additional yield from more nutrient-extractive crops would also accrue in terms of farm food availability and income, which could have positive feedbacks to soil fertility if inputs were purchased. Nutrient-extractive crops would transfer more nutrients to plant processes rather than soil processes. In most low-input systems, plant processes are much more subject to manipulation and management than are soil processes. Nutrients in residue, for example, may be used for livestock production, green manure, mulch, etc., whereas the soil is a less productive and riskier reservoir (i.e., more subject to fixation or erosion) for nutrient 'capital.' In many low-input systems in developing countries, soil resources are being rapidly degraded; in such systems to hold nutrients in the soil at the cost of reduced crop productivity may mean that those nutrients will be lost. It has been proposed that use of pigeon pea, a legume that is very efficient at extracting P from low-P soils, improves the overall fertility of cropping systems in low-P agroecosystems (Ae et al., 1990). The issue of 'soil mining' remains open, but there is as much reason to believe nutrient-extractive crops would enhance overall system productivity as there is to fear that they would degrade soil fertility.

An additional criticism of the value of nutrient-efficient crops in less developed countries is that such crops in some way retard the inevitable transition to high-input, high-yield agriculture. Nutrient-efficient crops that are responsive to nutrient inputs may actually encourage the use of fertility amendments by providing greater economic return to input use. Most conceivable mechanisms of nutrient efficiency would confer greater responsiveness to nutrient inputs from fertilizer as well as greater adaptation to unfertilized soil.

Another criticism of the value of nutrient-efficient crops in less developed countries is that they may encourage the exploitation of marginal lands that may better be left untouched, as they may be easily degraded and may have more value in nonagricultural use. The classic example of this is the forest margin in Latin America. Throughout Latin America, forests are first cleared for cultivation of subsistence crops by colonists, who after a few years move on, to be followed by cattlemen, in a process with serious ecological consequences for biodiversity, indigenous peoples, and the global climate. Would the introduction of nutrient-efficient crops accelerate this process? The reason that there are still large tracts of virgin forest in the tropics is that tropical forests are not comfortable places for civilized settlement;

insects, disease, and a hard, lawless, primitive life are the lot of many colonists. They have left the relative comfort and security of more traditional agricultural communities in the cooler highlands or more temperate latitudes because the soil in the highlands cannot support continuous erosion and sustained population growth. They move on to new clearings every few years when the soil fertility released from forest clearing and burning is exhausted. They are moving in search of soil fertility. By maintaining or improving crop productivity despite low soil fertility, the need to move and clear would be decreased. Nutrient-efficient crops would not increase the 'pull' of people into marginal lands; rather, they may decrease the 'push' of people away from their existing homes.

Phosphorus efficiency may be particularly important for less developed countries in the humid tropics and subtropics, where many soils have low P availability and poor responsiveness to P inputs (Sanchez and Uehara, 1980). Nitrogen can be fixed from the air by legumes or through the Haber process, but high-grade P rock deposits are limited to just a few nations (USA, Russia, and Morocco account for about 75% of world exports; Cathcart, 1980) and is therefore expensive to refine or import.

Modern Agriculture in the Aggregate

In very simplified terms, there are two types of cropping systems with respect to nutrient efficiency: high-input and low-input. In high-input systems, characteristic of the wealthy nations and some sectors of less developed countries, the primary value of nutrient-efficient crops is in reducing environmental contamination from excessive fertilization. In low-input systems, characteristic of less developed countries, the primary value of nutrient-efficient crops is in improving crop productivity.

Of the two types of systems, the need for nutrient-efficient crops in low input systems is most compelling. Eighty-five percent of the earth's population is in less developed countries. About half of the global workforce is engaged in agriculture in less developed countries. The outlook for food security in these nations is worrisome. Large numbers of disadvantaged people, especially children, are already malnourished. For example, in a poor nation such as Bangladesh, 84% of the children under age 5 are malnourished; even in a 'middle income' country such as Ecuador, 45% of the children under age 5 are malnourished (World

Bank, 1996). For the first time this century, world grain production per capita is on a decidedly downward trend (Brown and Kane, 1994). Each year adds another 90 million to the earth's population, yet yield potential in the main food crops has stagnated. The Worldwatch Institute has identified the lack of response to fertilization as one of three principle threats to future food security (Brown and Kane, 1994). Nutrient-efficient crops therefore have great potential value in addressing one of the central challenges facing humanity at this point in its history: how to sustainably feed itself.

OPPORTUNITIES FOR IMPROVING CROP NUTRIENT EFFICIENCY

The improvement of crop nutrient efficiency has been technically challenging. Traits controlling nutrient acquisition and utilization are highly complex and poorly understood. Studies in controlled environments may be artifactual, yet field trials are subject to large spatial and temporal variability. However, recent developments in our understanding of key processes, and our ability to exploit complex genetic traits have improved prospects for enhancing crop nutrient efficiency.

New methods to analyze and manipulate DNA should prove very useful in understanding and utilizing genetic variation for complex traits. The use of DNA markers will be helpful in assessing genetic diversity for nutrient efficiency traits, managing crop germplasm, selecting appropriate parents, evaluating crosses, and cloning relevant genes (Lee, 1995). Molecular markers will be especially useful for selection of genes for nutrient efficiency, considering the difficulties of using yield trials under field conditions to identify nutrient-efficient genotypes (Sussman and Gabelman, 1989; Lynch and Beebe, 1995). Many traits that should be linked with nutrient efficiency, such as root growth and architecture, are probably typical quantitative traits that exhibit substantial genotype $\times$ environment interaction, thereby complicating normal field selection programs. Theoretically, molecular markers should be useful in circumventing these problems by permitting direct genome screening (Tanksley et al., 1989), although inconsistent field performance over sites and seasons will make genetic markers less precise. However, if nutrient efficiency can be resolved into physiological mechanisms governed by discrete traits, then these traits could be tagged with molecular markers more reliably than could nutrient efficiency be measured as a quantitative trait by yield trials (Lynch and Beebe, 1995).

Advances in molecular biology have also enabled great strides in our understanding of the molecular basis for nutrient uptake and metabolism in plants (e.g., Smith et al., 1995). The artful use of mutants of Arabidopsis, tobacco, and other model plant systems has also presented important opportunities to critically test longstanding hypotheses concerning the adaptive value of specific traits or processes in conferring nutrient efficiency. For example, phosphatases are produced by plants under P stress and are thought to improve acquisition of P from soil organic P pools (Helal, 1990). Arabidopsis mutants lacking this function have been generated (Trull et al., 1996) so that this hypothesis can now be critically evaluated. Phosphatase is also an example of a relatively simple trait (a single gene product) that may be transferred or enhanced in other species through genetic engineering approaches. For efficiency traits that are related to transport processes at the level of cells and tissues, information about the genes and proteins responsible for nutrient transport may be very useful over the next 5 or 10 years. Traits related to membrane transport processes may be particularly important for K and Mg efficiency (Clark, 1975; Glass, 1980; Glass, 1989).

The biology of roots and root systems is receiving renewed attention by plant researchers. Our ability to observe roots has been improved with new techniques in tomography (Anderson and Hopmans, 1994) and minirhizotron techniques (Taylor, 1987). Our ability to analyze root form and function is improving through new quantitative tools such as kinematics (Silk, 1984), geometric simulation modeling (Lynch and Nielsen, 1996), fractal geometry (Fitter and Stickland, 1992; Nielsen, Lynch, and Weiss, 1997), and cost-benefit analysis (Bloom, Chapin, and Mooney, 1985). Substantial progress has been made in understanding the physiological and genetic basis of root tolerance of some important soil stresses such as Al toxicity (Kochian, 1995), drought (Spollen et al., 1993) and waterlogging (Drew and Stolzy, 1996). Genetic variation is increasingly being used to understand the impact of specific root traits among intraspecific comparisons (e.g., Bonser, Lynch, and Snapp, 1996). These methodological developments should enable new insights and interest in root biology over the coming decade. These will be critical to the development of more nutrient-extractive crops. Root architectural traits may be particularly important for P efficiency, since P is relatively immobile and heterogenously distributed in most soils (Atkinson, 1990). Efficient N scavenging would also be sensitive to root deployment in time and space.

Substantial progress is being made in our understanding of rhizosphere processes. The exudation of nutrient-mobilizing compounds into the rhizosphere, signaling between roots and symbionts such as mycorrhizal fungi and N-fixing bacteria, and the interaction of roots and the complex microbial communities in the rhizosphere are all critical areas of active research (Lynch, 1990). Enhanced understanding of these processes will be important in improving efficiency of micronutrient acquisition in many soils, with Fe as a leading example (Marschner and Römheld, 1994). Prospects for enhancing associative N fixation in the rhizosphere and from endophytes are promising (Boddey et al., 1995). Mycorrhizal associations are very important for nutrient efficiency, and are receiving considerable research attention by ecologists, and increasingly by agriculturists (Allen, 1992). When native soil populations have been suppressed through flooding or fumigation, inoculation with VAM fungi may be beneficial, but otherwise the practical application of mycorrhizal manipulation is uncertain, since the symbiosis is already ubiquitous in crop systems and the symbiosis is promiscuous. Genetic manipulation of crops to improve rhizosphere conditions is a very promising avenue for nutrient efficiency. For example, pigeon pea produces a low-molecular-weight organic acid that liberates P from Fe-P complexes that are not available to most other species (Ae et al., 1990). If the genes for this trait could be introduced to other species, their P acquisition efficiency in low-input environments may be substantially improved.

To better understand how nutrient efficiency in crops would impact nutrient cycling and sustained productivity, the incorporation of nutrient availability modules in important crop simulation models is also an important development that is beginning to occur.

CHALLENGES TO IMPROVING CROP NUTRIENT EFFICIENCY

I have argued that nutrient efficiency has an important role in modern agriculture, and that recent advances in several fields present new opportunities for improving crop nutrient efficiency. There are also several substantial obstacles to progress in this field that may limit the development of nutrient-efficient crops in the near future. Nutrient efficiency in crops, and indeed plant nutrition as a discipline, receives relatively low priority in most research and development agendas, and this situation appears to be worsening. Given the prevalence of low fertility soils around the world, and the economic and ecological limitations to intensive fertilization as a sustainable solution to integrated nutrient management, this is a puzzling situation. I propose that there are several interrelated structural features of the contemporary agricultural research enterprise that contribute to the low priority given to nutrient efficiency of crops.

Many agricultural scientists have poor awareness of nutritional constraints to crop production. University training is typically compartmentalized so that soil scientists do not learn much about crop physiology, and crop scientists do not learn much about soils. Where soils are treated in agronomy classes, the focus is usually on fertility management through fertilizers (Tisdale, Nelson, and Beaton, 1985). The interface between crops and soils, plant nutrition, is not emphasized. It is increasingly rare for units in US land-grant universities to have even one faculty member with this specialization. In contrast, it is common for agricultural colleges to have a whole department devoted to Plant Pathology and another devoted to Entomology. Much of classical plant breeding is for pest and disease resistance (Briggs and Knowles, 1967). As a result, most of the graduates from US agricultural colleges have a much greater understanding of, and appreciation for, pests and diseases as crop production constraints than for plant nutrition. This also reflects a geographic bias, as soil fertility has historically not been as problematic in Western Europe and the US as it is in most of the rest of the world. This shortcoming is particularly apparent in the humid tropics, where breeders and agronomists trained in the US often have very little awareness of the severe nutritional challenges posed by their weathered soils. This lack of awareness is reflected in low prioritization of work on nutrient efficiency. Notable exceptions to this generalization are Brazil, which has made considerable progress in adapting crops to their acid, low fertility soils, and Australia, where low soil fertility has historically been the impetus for outstanding research in plant nutrition.

A related set of constraints arise from reduced funding for international agricultural research. The CGIAR network of international research centers (including CIMMYT, CIAT, IRRI, ICRISAT, etc.) is in crisis because of prolonged, crippling budget cuts (Blake et al., 1994; Tribe, 1994). The decline in funding for international agricultural development has undercut a primary constituency for nutrient efficiency research, since the most compelling need for nutrient-efficient crops is in developing nations.

The rise of biotechnology and molecular biology has posed challenges as well as opportunities for nutrient efficiency research. These techniques are most powerful when used to manipulate single gene traits whose biology is already well understood. Most of the genetically engineered crops now being developed incorporate single gene traits conferring resistance to certain pesticides, insects, or diseases (Rissler and Mellon, 1993). In contrast, crop nutrient efficiency may involve multigenic traits whose biology is not well understood. Many of the successful crop genetic engineering projects have received substantial funding from the private sector, which is not motivated to pursue technology with diffuse public benefits, such as nutrient efficiency (Busch et al., 1991).

Although current research in crop nutrient efficiency lags behind efforts directed to other biotic and abiotic stresses, the magnitude of soil fertility problems in modern agriculture suggest that interest in crop nutrient efficiency will continue. For example, the international agricultural development community is now granting renewed attention to the fact that low P availability is a primary constraint to crop productivity in Africa (ASA, 1996). Although much of this renewed interest is directed to fertilizers, eventually the role of nutrient-efficient crop genotypes will be noticed by economists and policymakers. Over the long term, increased population pressures on agricultural land, combined with the limited availability of high-grade P ore deposits, will make P efficiency, at least, an important attribute of future crops (Cathcart, 1980).

REFERENCES

Ae, N., J. Arihara, K. Okada, T. Yoshihara, and C. Johansen. (1990). Phosphorus uptake by pigeon pea and its role in cropping systems of the Indian subcontinent. *Science* 248: 477-480.

Allen, M.F. (1992). *Mycorrhizal Functioning: An Integrative Plant-Fungal Process.* New York, USA: Chapman and Hall.

Anderson, S.H. and J.W. Hopmans, eds. (1994). *Tomography of Soil-Water-Root Processes.* SSSA Special Publication 36. Madison, Wisconsin, USA: American Society of Agronomy, Soil Science Society of America.

Anonymous. (1996). *Packers Fresh Trends-1996.* Chicago, Illinois, USA: Vance Publishing.

ASA (Agronomy Society of America). (1996). *Agronomy Abstracts.*

Atkinson, D. (1990). Influence of root system morphology and development on the need for fertilizers and the efficiency of use. In *Crops as Enhancers of Nutrient Use*, eds. V.C. Baligar and R.R. Duncan. San Diego, California, USA: Academic Press, pp. 411-451.

Batten, G.D. (1992). A review of phosphorus efficiency in wheat. *Plant and Soil* 146: 163-168.

Blair, G. (1993). Nutrient efficiency-what do we really mean? *Developments in Plant and Soil Science* 50: 205-213.

Blake, R.O., D.E. Bell, J.T. Mathews, R.S. McNamara, M.P. McPherson, and M. Yudelman. (1994). *Feeding 10 Billion People in 2050: The Key Role of the CGIARs International Agricultural Research Centers*. Washington, D.C., USA: Committee on Agricultural Sustainability for Developing Countries.

Bloom, A.J., F.S. Chapin III, and H.A. Mooney. (1985). Resource limitation in plants-an economic analogy. *Annual Review of Ecology and Systematics* 16: 363-392.

Boddey, R.M., O.C. de Oliveira, S. Urquiaga, V.M. Reis, F.L. de Olivares, V.L.D. Baldani, and J. Dobereiner. (1995). Biological nitrogen fixation associated with sugar cane and rice: contributions and prospects for improvement. *Plant and Soil* 174: 195-209.

Bonser, A.M., J. Lynch, and S. Snapp. (1996). Effect of phosphorus deficiency on growth angle of basal roots of *Phaseolus vulgaris* L. *The New Phytologist* 132: 281-288.

Borlaug, N.E. (1972). *The green revolution, peace, and humanity*. Speech delivered upon receipt of the 1970 Nobel Peace Prize. CIMMYT reprint and translation series No. 3. El Batan, Mexico: Centro Internacional de Mejoramiento de Maiz y Trigo.

Briggs, F.N. and P.F. Knowles. (1967). *Introduction to Plant Breeding*. New York, USA: Reinhold Publishing Corporation.

Brown, J.C. and V.D. Jolley. (1989). Plant metabolic responses to iron-deficiency stress. *Bioscience* 39: 546-551.

Brown, L.R. and H. Kane. (1994). *Full House: Reassessing the Earth's Population Carrying Capacity*. New York, USA: Norton and Co.

Busch, L., W.B. Lacy, J. Burkhardt, and L.R. Lacy. (1991). *Plants, Power, and Profit: Social, Economic, and Ethical Consequences of the New Biotechnologies*. Cambridge, UK: Blackwell.

Callaway, M.B. and C.A. Francis. (1993). *Crop Improvement for Sustainable Agriculture*. Lincoln, Nebraska, USA: University of Nebraska Press.

CAST (Council for Agricultural Science and Technology). (1985). *Agriculture and Ground Water Quality*. Council for Agricultural Science and Technology Report 103.

Cathcart, J.B. (1980). World phosphate reserves and resources. In *The Role of Phosphorus in Agriculture*, eds. F.E. Khasawneh, E.C. Sample, and E.J. Kamprath. Madison, Wisconsin, USA: American Society of Agronomy, Crop Science Society of America, Soil Science Society of America, pp. 1-18.

Clark, R.B. (1975). Differential magnesium efficiency in corn inbreds. I. Dry matter yield and mineral element composition. *Soil Science Society of America Proceedings* 39: 488-491.

Clark, R.B. (1990). Physiology of cereals for mineral nutrient uptake, use, and efficiency. In *Crops as Enhancers of Nutrient Use*, eds. V.C. Baligar and R.R. Duncan. San Diego, California, USA: Academic Press, pp. 131-209.

Cock, J. (1985). *Cassava: New Potential for a Neglected Crop*. Boulder, Colorado, USA: Westview Press.

Drew, M.C. and L.H. Stolzy. (1996). Growth under oxygen stress. In *Plant Roots: The Hidden Half, 2nd edition*, eds. Y. Waisel, A. Eshel, and U. Kafkafi. New York, USA: Marcel Dekker, pp. 397-414.

FAO (Food and Agriculture Organization). (1996a). *FAO Production Yearbook*. Rome, Italy: FAO.

FAO (Food and Agriculture Organization). (1996b). *FAO Fertilizer Yearbook*. Rome, Italy: FAO.

Feil, B., R. Thiraporn, and P. Stamp. (1992). Can maize cultivars with low mineral nutrient concentrations in the grains help to reduce the need for fertilizers in third world countries? *Plant and Soil* 146: 227-231.

Fitter, A.H. (1991). Characteristics and functions of root systems. In *Plant Roots: The Hidden Half*, eds. Y. Waisel, A. Eshel, and U. Kafkafi. New York, USA: Marcel Dekker, pp. 3-24.

Fitter, A.H. and Stickland T.R. (1992). Fractal characterization of root system architecture. *Functional Ecology* 6: 632-635.

Fitter, A.H., T.R. Stickland, M.L. Harvey and G.W. Wilson. (1991). Architectural analysis of plant root systems. I. Architectural correlates of exploitation efficiency. *The New Phytologist* 118: 375-382.

Follett, R.F. and D.J. Walker. (1989). Ground water quality concerns about nitrogen. In *Nitrogen Management and Ground Water Protection*, ed. R.F. Follett. New York, USA: Elsevier, pp. 1-22.

Fraser, P. and C. Chilvers. (1981). Health aspects of nitrate in drinking water. *The Science of the Total Environment* 18: 103-116.

Glass, A.D.M. (1980). Varietal differences in potassium uptake by barley [Selection tests with Canadian cultivars]. *Plant Physiology* 65: 160-164.

Glass, A.D.M. (1989). Physiological mechanisms involved with genotypic differences in ion absorption and utilization. *HortScience* 24: 559-564.

Graham, R.D. (1984). Breeding for nutritional characteristics in cereals. In *Advances in Plant Nutrition, Volume 1*, eds. P.B. Tinker and A. Läuchli, New York, USA: Praeger, pp. 57-102.

Graham, R.D., J.S. Ascher, and S.C. Hynes. (1992). Selecting zinc-efficient cereal genotypes for soils of low zinc status. *Plant and Soil* 146: 241-250.

Gutschick, V.P. (1993). Nutrient-limited growth rates: roles of nutrient efficiency and of adaptations to increase uptake rates. *Journal of Experimental Botany* 44: 41-52.

Hallberg, G. (1989). Nitrate in ground water in the United States. In *Nitrogen Management and Ground Water Protection*, ed. R.F. Follett. New York, USA: Elsevier, pp. 35-74.

Hecht, S.B., ed. (1982). *Amazonia: Agriculture and Land Use Research; Proceedings*. Cali, Colombia: Centro Internacional de Agricultura Tropical.

Helal, H.M. (1990). Varietal differences in root phosphatase activity as related to the utilization of organic phosphates. *Plant and Soil* 123: 161-163.

Jürgens-Gschwind S. (1989). Ground water nitrates in other developed countries (Europe)–relationships to land use patterns. In *Nitrogen Management and Ground Water Protection*, ed. R.F. Follett. New York, USA: Elsevier, pp. 75-138.

Kochian, L.V. (1995). Cellular mechanisms of aluminum toxicity and resistance in plants. *Annual Review of Plant Physiology and Plant Molecular Biology* 46: 237-260.

Kohli, M.M. and S. Rajaram. (1988). *Wheat Breeding for Acid Soils: Review of Brazilian/CIMMYT Collaboration, 1974-1986.* Mexico, D.F., Mexico: Centro Internacional de Mejoramiento de Maiz y Trigo.

Lal, R. (1985). Soil erosion and its relation to productivity in tropical soils. In *Soil Erosion and Conservation*, eds. S.A. El-Swaify, W.C. Moldenhauer, and A. Lo. Madison, Wisconsin, USA: American Society of Agronomy, pp. 237-247.

Lee, M. (1995). DNA markers and plant breeding programs. *Advances in Agronomy* 55: 265-344.

Louisiana State University. (1995). *Nutrient-Enhanced Coastal Ocean Productivity.* Proceedings of 1994 Synthesis Workshop, Louisiana Sea Grant College Program. Baton Rouge, Louisiana, USA: Louisiana State University.

Lynch, J. (1995). Root architecture and plant productivity. *Plant Physiology* 109: 7-13.

Lynch, J. and A. González. (1993). Canopy nutrient allocation in relation to incident light in the tropical fruit tree *Borojoa patinoi. Journal of the American Society of Horticultural Science* 118: 777-785.

Lynch, J. and K.L. Nielsen. (1996). Simulation of root system architecture. In *Plant Roots: The Hidden Half, 2nd edition*, eds. Y. Waisel, A. Eshel, and U. Kafkafi. New York, USA: Marcel Dekker, pp. 247-257.

Lynch, J. and N. Rodriguez. (1994). Photosynthetic nitrogen use efficiency in relation to leaf longevity in common bean. *Crop Science* 34: 1284-1290.

Lynch, J.M. (1990). *The Rhizosphere.* New York, USA: John Wiley & Sons.

Lynch, J.P. and S.E. Beebe. (1995). Adaptation of beans to low soil phosphorus availability. *HortScience* 30: 1165-1171.

Marschner, H. (1995). *Mineral Nutrition of Higher Plants*, 2nd ed. London, UK: Academic Press.

Marschner, H. and V. Römheld. (1994). Strategies of plants for acquisition of iron. *Plant and Soil* 165: 261-274.

Nielsen, K., J.P. Lynch, and H.N. Weiss. (1997). Fractal geometry of bean root systems. Correlations between spatial and fractal dimension. *American Journal of Botany* 84: 26-33.

NRC (National Research Council). (1989). *Alternative Agriculture.* Washington, D.C., USA: National Academy Press.

Rissler, J. and M. Mellon. (1993). *Perils Amidst the Promise: Ecological Risks of Transgenic Crops in a Global Market.* Cambridge, Massachusetts, USA: Union of Concerned Scientists.

Ryden, J.C., J.K. Syers, and R.F. Harris. (1973). Phosphorus in runoff and streams. *Advances in Agronomy* 25: 1-45.

Sample, E.C., R.J. Soper, and G.J. Racz. (1980). Reactions of phosphate fertilizers in soils. In *The Role of Phosphorus in Agriculture*, eds. F.E. Khasawneh, E.C. Sample, and E.J. Kamprath. Madison, Wisconsin, USA: American Society of Agronomy, Crop Science Society of America, Soil Science Society of America, pp. 263-310.

Sanchez, P.A. (1976). *Properties and Management of Soils in the Tropics.* New York, USA: John Wiley & Sons.

Sanchez, P.A. and S.W. Buol. (1975). Soils of the tropics and the world food crisis. In *Food: Politics Economics Nutrition and Research*, ed. P.H. Abelson. Washington, D.C., USA: American Association for the Advancement of Science, pp. 115-120.

Sanchez, P.A. and G. Uehara. (1980). Management considerations for acid soils with high phosphorus fixation capacity. In *The Role of Phosphorus in Agriculture, eds.* F.E. Khasawneh, E.C. Sample, and E.J. Kamprath. Madison, Wisconsin, USA: American Society of Agronomy, Crop Science Society of America, Soil Science Society of America, pp. 471-514.

Schwartz, H.F. and M.A. Pastor-Corralles, eds. (1989). *Bean Production Problems in the Tropics.* 2nd ed. Cali, Colombia: Centro Internacional de Agricultura Tropical.

Shiva, V. (1991). The green revolution in the Punjab. *The Ecologist* 21: 57-60.

Silk, W.K. (1984). Quantitative descriptions of development. *Annual Review of Plant Physiology* 35: 479-518.

Smith, F.W., P.M. Ealing, M.J. Hawkesford, and D.T. Clarkson. (1995). Plant members of a family of sulfate transporters reveal functional subtypes. *Proceedings of the National Academy of Sciences of the USA* 92: 9373-9377.

Spollen, W.G., R.E. Sharp, I.N. Saab, and Y. Wu. (1993). Regulation of cell expansion in roots and shoots at low water potentials. In *Water Deficits: Plant Responses from Cell to Community*, eds. J.A.C. Smith and H. Griffiths. Oxford, UK: BIOS Scientific Publishers, pp. 37-52.

Sussman, M.R. and W.H. Gabelman. (1989). Genetic aspects of mineral nutrition: future challenges and directions. *HortScience* 24: 591-594.

Tanksley, S.D., N.O. Young, A.H. Paterson, and M.W. Bonierbale. (1989). RFLP mapping in plant breeding: new tools for an old science. *Bio/Technology* 7: 257-264.

Taylor, H.M. (1987). *Minirhizotron Observation Tubes: Methods and Applications for Measuring Rhizosphere Dynamics.* ASA Special Publication 50. Madison, Wisconsin, USA: American Society of Agronomy.

Thung, M. (1990). Phosphorus: a limiting nutrient in bean (*Phaseolus vulgaris* L.) production in Latin America and field screening for efficiency and response. In *Genetic Aspects of Plant Nutrition*, eds. N. El Bassam, M. Damrouth, and B.C. Loughman. Dordrecht, The Netherlands: Kluwer Academic Publishers, pp. 501-521.

Tisdale, S.L., W.L. Nelson, J.D. Beaton. (1985). *Soil Fertility and Fertilizers, 4th edition.* New York, USA: Macmillan Publishing Co.

Tribe, D. (1994). *Feeding and Greening the World: The Role of International Agricultural Research.* Wallingford, UK: CAB International.

Trull, M.C., M.J. Guiltinan, J.P. Lynch, and J. Deikman. (1996). Characterization of phosphatase underproducing (PUP) mutants of *Arabidopsis. Plant Physiology* 111 Suppl., abstract 126.

Uehara, G. and J. Keng. (1975). Management implications of soil mineralogy in Latin America. In *Soil Management in Tropical America*, eds. E. Bornemisza and A. Alvarado. Raleigh, North Carolina, USA: North Carolina State University, pp. 351-363.

Whiteaker, G., G.C. Gerloff, W.H. Gabelman, and D. Lindgren. (1976). Intraspecific differences in growth of beans at stress levels of phosphorus. *Journal of the American Society of Horticultural Science* 101: 472-475.

World Bank. (1996). *From Plan to Market: World Development Report, 1996.* New York, USA: Oxford University Press.

Wortmann, C.S. and D.J. Allen. (1994). *African Bean Production Environments: Their Definition, Characteristics and Constraints.* Occasional Paper Series No. 11. Dar es Salaam, Tanzania: Network on Bean Research in Africa.

SUBMITTED: 03/20/97
ACCEPTED: 07/11/97

Index

Acid sulfate soils, 46
Agricultural policy, 31-36,48
Aluminum, 146,148-149,151
  tolerant genotypes, 251
Ammonium fertilizers, 83,86,88,93,
  96,100
Ammonia volatilization, 87-89,
  91-92,96

Boron, 182
  availability, 182-183
  B × pH interaction, 183-184
Burning stubble, 182

Calcium, 17,89
Cate-Nelson graphical method, 61
Cation exchange capacity, 57,88,97,169
Cation saturation ratio concept, 62
Copper, 172
  availability, 173-174
Critical concentration, 63,67,69
Crop nutrient efficiency, 242
  acquisition, 249,252
  adaptation, 245
  genotypic differences, 245
  mycorrhiza, 257
  P, 254
  responsiveness, 245
  utilization, 245

Desertification, 43
Diagnostic criteria, 62-63
Disease control, 118
DRIS, 64-65
Drought stress, 223-224

Eutrophication, 249
Extractants, 57-58,60,67,69

Fallowing, 40
Fertilizers, 8-10,12,15,20,22
  banding, 71,83,101,150,167,
    175-176
  fertilizer N, 10-11,13,16-17,38,
    71,82,104
  fertilizer P, 11,17-18,71
  fertilizer K, 11,19,71,101
  fertilizer demand by region, 21
  recovery, 82
  slow-release, 100
  sulfur-coated urea, 83,92,99
Fiber, 7-8,20
Foliar analyses, 59,64,70
Food-deficit countries, 7,23
Food production, 5-8,10-11,19-20,23
Food imports, 5
Francolite, 143

Glaucousness, 224
Global warming, 93
Grain legumes, 127
Green manure, 132-133
  Azolla, 132-133
  Anabaena, 132
Green revolution, 6

Harvest index, 223
High-yielding varieties, 9,11,16,20,
  22
Hunger, 7

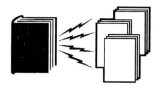

Haworth
DOCUMENT DELIVERY
SERVICE

This valuable service provides a single-article order form for any article from a Haworth journal.

- *Time Saving:* No running around from library to library to find a specific article.
- *Cost Effective:* All costs are kept down to a minimum.
- *Fast Delivery:* Choose from several options, including same-day FAX.
- *No Copyright Hassles:* You will be supplied by the original publisher.
- *Easy Payment:* Choose from several easy payment methods.

Open Accounts Welcome for . . .
- Library Interlibrary Loan Departments
- Library Network/Consortia Wishing to Provide Single-Article Services
- Indexing/Abstracting Services with Single Article Provision Services
- Document Provision Brokers and Freelance Information Service Providers

---

MAIL or *FAX* THIS ENTIRE ORDER FORM TO:

Haworth Document Delivery Service
The Haworth Press, Inc.
10 Alice Street
Binghamton, NY 13904-1580

or FAX: 1-800-895-0582
or CALL: 1-800-429-6784
9am-5pm EST

PLEASE SEND ME PHOTOCOPIES OF THE FOLLOWING SINGLE ARTICLES:

1) Journal Title: _____
   Vol/Issue/Year: _____ Starting & Ending Pages: _____
   Article Title: _____
   _____

2) Journal Title: _____
   Vol/Issue/Year: _____ Starting & Ending Pages: _____
   Article Title: _____
   _____

3) Journal Title: _____
   Vol/Issue/Year: _____ Starting & Ending Pages: _____
   Article Title: _____
   _____

4) Journal Title: _____
   Vol/Issue/Year: _____ Starting & Ending Pages: _____
   Article Title: _____
   _____

(See other side for Costs and Payment Information)

COSTS: Please figure your cost to order quality copies of an article.

1. Set-up charge per article: $8.00
   ($8.00 × number of separate articles)   _____

2. Photocopying charge for each article:

   1-10 pages: $1.00   _____

   11-19 pages: $3.00   _____

   20-29 pages: $5.00   _____

   30+ pages: $2.00/10 pages   _____

3. Flexicover (optional): $2.00/article   _____

4. Postage & Handling: US: $1.00 for the first article/

   $.50 each additional article   _____

   Federal Express: $25.00   _____

   Outside US: $2.00 for first article/

   $.50 each additional article _____

5. Same-day FAX service: $.50 per page   _____

GRAND TOTAL: _____

METHOD OF PAYMENT: (please check one)

❑ Check enclosed ❑ Please ship and bill. PO # _____
(sorry we can ship and bill to bookstores only! All others must pre-pay)

❑ Charge to my credit card: ❑ Visa; ❑ MasterCard; ❑ Discover;
❑ American Express;

Account Number:_____ Expiration date:_____

Signature: **X** _____

Name: _____ Institution: _____

Address: _____

_____

City: _____ State:_____ Zip:_____

Phone Number: _____ FAX Number: _____

MAIL or *FAX* THIS ENTIRE ORDER FORM TO:

Haworth Document Delivery Service | **or FAX:** 1-800-895-0582
The Haworth Press, Inc. | **or CALL:** 1-800-429-6784
10 Alice Street | (9am-5pm EST)
Binghamton, NY 13904-1580 |